UNDERSTANDING THE

3   570.282 GOL

UNDERSTANDING THE LIGHT MICROSCOPE

A Computer-aided Introduction

D.J. Goldstein

Department of Biomedical Science,
University of Sheffield,
Sheffield S10 2TN, U.K.
email address: d.j.goldstein@sheffield.ac.uk

ACADEMIC PRESS

San Diego   London   Boston   New York
Sydney   Tokyo   Toronto

Copyright © 1999 by ACADEMIC PRESS

Academic Press
24–28 Oval Road, London NW1 7DX, UK
http://www.hbuk.co.uk/ap/

Academic Press
A Harcourt Science and Technology Company
525 B Street, Suite 1900, San Diego, California 92101-4495, USA
http://www.apnet.com

ISBN 0-12-288660-7

A catalogue record for this book is available from the British Library

Library of Congress Catalog Card Number: 99-61700

Printed in Great Britain by MPG Books Ltd, Bodmin, Cornwall

99 00 01 02 03 MP 9 8 7 6 5 4 3 2 1

Contents

Diagrams in Text

OTHER DIAGRAMS

The following diagrams can be viewed with the aid of the computer programs.

The two diagrams with the prefix K are the transmitted-light and phase-contrast ray diagrams of the interactive Kohler program, and include the light source, lamp lens, field stop, condenser, object, objective, eyepiece and image. The size and position of the iris diaphragms or phase annulus and the focus of the lenses are all alterable by the user, and result in corresponding changes in the ray paths and the image.

K1   Köhler illumination (bright-field)
K2   Köhler illumination (phase-contrast)

Static diagrams with the prefix Z can be viewed while the Zernike program is running by pressing function key F4. Diagrams Z1–Z14 inclusive show various imaging modes, while Z15–Z18 illustrate aspects of the theory of diffraction.

Z1   Coherent Illumination
Z2   Oblique Coherent Illumination
Z3   Partially Coherent Illumination ('critical')
Z4   Partially Coherent Illumination (Köhler)
Z5   Incoherent Illumination
Z6   Phase Contrast
Z7   Interference Contrast (differential)
Z8   Interference Contrast (Jamin–Lebedeff)
Z9   Schlieren/Peripheral Dark-field
Z10   Apodization/Central Dark-field
Z11   Hoffman's 'Modulation Contrast'
Z12   'Gegenfeld'

Diagrams with the prefix E are generated by the program when suggested exercises (Chapter 17) are carried out by the user.

Preface

This book and the accompanying computer programs owe their origin to an article on phase contrast microscopy which I wrote a good many years ago. Shortly after the article appeared in 1982 I realised, with some embarrassment, that I had been confused by the subject and was still puzzled by it. Puzzlement is, however, better than confusion, since the recognition of ignorance is an essential if painful step on the road to knowledge. In order to understand how a transparent microscopic object affects the phase of the light diffracted by it I wrote a short computer program for a BBC model B microcomputer which had recently become available to me. This program employed a somewhat mysterious algorithm called the fast Fourier transform which was given to me by a kind colleague,[1] and met my immediate need. It later occurred to me that the program could (fairly) easily be extended to calculate the effect on bright-field images of a finite objective aperture, and over months and then years the Zernike program, named after the Dutch physicist,[2] gradually grew to include many other aspects of light microscopy. It was originally written in interpreted BBC BASIC but was later converted to compiled TurboBasic for IBM-compatible computers.

Computers can imitate various aspects of microscopic imaging (see e.g. Galbraith and Sanderson, 1980; Galbraith, 1982; Krakow, 1984, 1991; Spycher et al., 1988), and sophisticated computer programs are routinely used by optical engineers and lens designers. The Zernike program is rather different. It is a relatively simple program based on elementary diffraction theory (Goldstein, 1991a), which runs happily on a small

---

[1] Dr. J. Bennett of the Department of Electronic and Electrical Engineering, University of Sheffield.
[2] Frits Zernike (1888–1966) of Groningen (Netherlands). Inventor of the phase-contrast microscope, Nobel prize-winner 1953.

personal computer. It was written by a biologist who is neither a mathematician nor a computer expert, and although intended primarily as a learning aid for microscopists who want to understand their instrument and its limitations better, it has been used in research (e.g. Goldstein 1991b, 1992; Davidson and Butler, 1992). Aspects of image formation which can be simulated include the effects of the aperture, spherical aberration and focus of the objective lens, and the coherence of the illumination. In addition to ordinary bright-field illumination the program covers oblique illumination, phase-contrast, central and peripheral dark-field, Schlieren, 'modulation contrast', apodization, interference microscopy (both shearing and differential), fluorescence, and confocal scanning.

The Zernike program can also simulate various phenomena connected with the physics of light including Fraunhofer and Fresnel diffraction by slits and gratings, interference of light from two slits, and the formation of a bright region in the middle of the shadow of an opaque object. It may therefore be useful for students interested in general optics rather than microscopy.

Working on (and playing with) the program has cost me innumerable hours which should have been devoted to other matters, but also taught me things about microscopy and diffraction which were missing from my formal biological and medical education. I hope that the program will entertain my fellow microscopists and give them additional insight into the operation and limitations of their instrument. A little perseverance may be necessary for the less numerate, but what one non-mathematician has created should not be beyond the understanding of others!

Three other programs complete the set. Kohler[1] is a simple, interactive program intended to help one learn how to operate a bright-field or phase-contrast microscope correctly, while Nicol[2] simulates aspects of quantitative polarized light microscopy. In Kohler the settings of the microscope, and in Nicol the properties of the object can be made random to provide practice respectively in adjusting the microscope and analysing an unknown anisotropic specimen. Snellius[3] is a simple ray-tracing program which uses Snel's law of refraction to follow rays through a lens system, and demonstrates various aberrations which afflict real lenses.

After writing the program it became evident that an accompanying, explanatory handbook was essential – some existing textbooks were too

---

[1] August Karl Johann Valentin Köhler (1866–1948), microscopist of Carl Zeiss and the University of Jena.
[2] William Nicol (1768–1851), geologist of Edinburgh.
[3] Willebrord Snel (1580–1628), physicist, mathematician and astronomer of Leiden.

elementary, and I found others to be too difficult![1] Hence the present volume, which is intended to be not only a guide to the use of the programs but also an introduction to diffraction theory, especially as applicable to the microscope. Standard texts (e.g. Martin, 1966; Jenkins and White; 1957; Born and Wolf, 1975; Reynolds *et al.*, 1989; Pluta, 1988–9) should be consulted if a deeper understanding of the theoretical basis of microscopy and the physics of light is required. It is assumed that readers are already familiar with the elementary geometrical optics of the microscope (see e.g. Bradbury, 1989), but some comments on geometrical optics and on polarized light microscopy can be found in the chapters respectively on the Kohler, Snellius and Nicol programs.

If possible the handbook should be read with a computer and ideally also a real microscope to hand. Practical exercises illustrating various phenomena discussed in the book are described by several authors (e.g. Lacey, 1968).

Computer-literate readers who want to see how the programs work can examine the source code (text files with the suffix .BAS or .INC) using any word-processor. Suggestions for additions or alterations to the program would be greatly appreciated.

If you are intolerant of the printed word or want to start using the Zernike program as soon as possible, turn now to Chapter 16. This describes what the program does and how to use it. Additional information is available elsewhere in the book, and help is obtainable while the program is running by touching function key F1.

Readers more versed than I am in mathematics and physics will probably not learn much from this book, but I hope that they will derive some entertainment from the computer programs. I should be grateful if any significant mistakes in the book or programs are brought to my attention, preferably in a kindly manner via the publishers. My explanation for such (at present hypothetical) errors is identical with that given by Dr Johnson when asked by a lady why his dictionary defined 'pastern' as the 'knee of a horse': 'Ignorance, Madam, sheer ignorance'.

---

[1] The publishers of Martin's (1966) book claim that it '... is not necessary for his readers to have a greater knowledge of mathematics and physics than that considered essential to serious users of the microscope.' If this is true, very few biologists can be serious microscopists.

Chapter 1

Simple Ray Optics (the Kohler Program)

1.1 THE NEED FOR APPROXIMATIONS IN OPTICS

When explaining optical phenomena, intellectual rigour is sometimes unnecessary, undesirable, or even impossible. This may be due to the fact that no comprehensive theory of the nature of light seems to be comprehensible by ordinary microscopists. Light behaves like discrete particles (quanta) when it is produced from or transformed into other types of energy, but like a form of wave motion while it is travelling from one point to another. It is important in physics to use the appropriate model when dealing with a particular problem, and quantum theory and wave mechanics are largely incompatible with each other. Quantum theory is outside the scope of this book, but the wave nature of light must be taken into account when dealing with such topics as resolving power and the imaging of transparent objects, and microscopic image formation is dealt with in terms of waves in other chapters.

In the following discussion of the action of lenses, wave properties of light are however ignored and light is considered to travel in straight lines (rays). This simplification is possible since the wavelength of light is negligible compared with the extent of the transparent media under discussion – light waves become relevant only when dealing with objects comparable in size with the wavelength of light, or with the behaviour of light near edges. Simple ray diagrams such as Figure 1.2 adequately explain many important concepts in microscopy including reflection, refraction, how ideal ('thin') lenses work, focal length, magnification, erect and inverted images, field stops and aperture stops.

1.2   A PRELIMINARY NOTE ON SOME CONVENTIONS AND TERMINOLOGY

It is necessary to use an internally consistent system of conventions in optical equations and diagrams. In the convention followed here light travels in diagrams from left to right; distances to the left of a lens and the radii of surfaces concave to the left are negative, while distances to the right of a lens and the radii of lens surfaces concave to the right are positive. We shall deal only with perfectly centred lens systems with plane or spherical surfaces, symmetrical about the optic axis.

The *refractive index* of a medium is inversely proportional to the speed of light in the medium. The refractive index of a vacuum, and of air for most practical purposes, is unity.

The *poles* of a lens are the points where the surfaces of the lens intersect the optic axis.

The *magnification* M of a lens equals h'/h, where h' and h are respectively the heights of the image and the object. This definition refers to the *lateral* magnification of a flat object perpendicular to the optic axis – a three-dimensional object will however be imaged into a three-dimensional space, and the *longitudinal* magnification of such an image is in general rather different from the lateral one. Magnification is often indicated by a multiplication sign, e.g. a lens which under particular conditions gives a 10-fold enlargement of the image is called a × 10 lens.

The *optical tube length* of a microscope, conventionally 160 mm, is the distance from the back focal plane of the objective to the primary image plane at the front focal plane of the eyepiece. The *primary magnification* due to the objective equals the optical tube length divided by the focal length of the lens, while the magnification of the eyepiece equals the arbitrary distance of 250 mm divided by the focal length of the eyepiece. The *total* magnification of a compound microscope (ignoring factors such as binocular bodies etc.) equals the objective magnification multiplied by the eyepiece magnification. For example, with an objective of 4 mm focal length and an eyepiece of 25 mm focal length, the final magnification is $160/4 \times 250/25 = \times 400$.

1.3   REFLECTION AND REFRACTION OF RAYS

1.3.1   Reflection.   According to the familiar law of reflection, if a ray of light is reflected from a surface the angle of the incident light, relative to the normal to the surface, equals the angle of reflection. The incident ray, the normal and the reflected ray all lie in the same plane.

1.3.2   Refraction.   The law of refraction was formulated by Willebrord Snel, also known by the Latin version of his name (Snellius), and

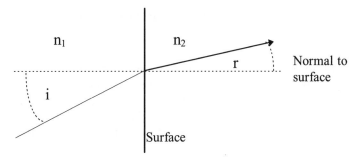

Figure 1.1 Snel's law of refraction

who sometimes appears in English texts as Snell. According to Snel, when a light ray passes from a homogeneous transparent medium of refractive index n_1 into a second medium of refractive index n_2, the angle of incidence i at the interface (relative to the normal to the surface) is related to the angle of refraction r by

$$\frac{\sin(i)}{\sin(r)} = \frac{n_2}{n_1}$$

Not all the light necessarily passes into the second medium. If T and R are respectively the intensities of the transmitted and reflected rays, and the ray strikes the surface at 'normal' incidence (perpendicularly), then approximately

$$T = \left(\frac{2n_1}{n_1 + n_2}\right)^2 \quad \text{and} \quad R = \left(\frac{n_2 - n_1}{n_2 + n_1}\right)^2$$

For an ordinary glass–air surface the intensity of the perpendicularly reflected light can be substantial, perhaps 4–5% of the total. In practice matters are often more complicated than this. As the angle of incidence increases so does the intensity of the reflected light, at the expense of the transmitted light. Except in the case of normal incidence the relative intensities of the reflected and refracted beams are also influenced by the state of polarization of the light. Reflection and refraction at a surface can also be markedly affected by interference with light reflected at other surfaces. Thus if the optical path through a thin transparent coating on a piece of glass is 0.25 wavelengths, light reflected off the second surface of the coating will be 0.5 wavelengths out of phase with and will tend to interact destructively with that reflected off the first surface. The result is that less light is reflected and more is transmitted; this is the basis of the antireflection coatings widely used in optics.

If light travelling in a medium of refractive index n_1 strikes the interface with a medium of lower refractive index n_2 at an angle i such that $\sin(i) \cdot n_2/n_1$ is greater than unity, all the light is reflected back into the first medium and none passes through the interface. This is *total internal reflection*.

1.3.3 Spherical lenses. Non-spherical lens surfaces can be made ('figured') by hand, but with few exceptions the surfaces of lenses used in microscopy are either plane or spherical, since such surfaces can be produced accurately, reproducibly and relatively inexpensively by automatic machinery. This imposes severe practical restrictions on the lens designer.

1.3.4 Gaussian optics. *Ray-tracing*, i.e. studying the passage of light rays through lenses using Snel's law, is in principle straight-forward. Ray tracing through a complicated three-dimensional system however involves tedious trigonometry, and although this is just what computers are good at (see Chapter 2 for a discussion of the 'Snellius' program), if algebraic methods are to be used it is expedient to simplify matters as far as possible. This is achieved in *Gaussian* optics by considering only *paraxial rays*, i.e. those near the optic axis. The angles of incidence and refraction of rays at lens surfaces are then necessarily small, and if measured in radians can be taken to be equal to their sines. For the mathematically inclined this is equivalent to saying that the sine of a small angle A can be adequately represented by the first term only of the expansion

$$\sin(A) = A - \frac{A^3}{3!} + \frac{A^5}{5!} - \cdots$$

Gaussian optics are widely used in describing and analysing the behaviour and aberrations of lenses, and detailed descriptions can be found in standard texts (e.g. Martin, 1966; Born and Wolf, 1975). Fortunately for us, for many purposes in microscopy matters can be simplified yet further.

1.3.5 The 'thin lens' approximation. In this special case of Gaussian optics it is assumed that the thickness of a lens (i.e. the distance between its poles) is negligibly small compared either with the radii of curvature of the lens surfaces, or with the distances from the lens to the object and to the image. Under these conditions it is possible to derive some simple but useful relationships which can be found in most optical texts but are given here for convenience. A discussion of how real ('thick') lenses behave can be found in Chapter 2.

If rays from a distant object travel parallel to the optic axis, they are brought to a focus by a thin converging lens at a point on the optic axis called the *back focal point*, which lies at a distance from the lens called the *focal length*. More generally, the focal length f of a thin lens is given by the basic *lens formula*

$$\frac{1}{f} = \frac{1}{v} - \frac{1}{u}$$

where u and v are respectively the distances of the object and image from the lens. Note that in our convention the numerical value of $-1/u$ is positive if u lies to the left of the lens. If the object is at an infinite (or at least a very great) distance from the lens, $1/u$ tends to zero and the equation reduces to

$$f = v.$$

The *back focal plane* is perpendicular to the optic axis and passes through the back focal point. The front focal point and plane lie at a distance in front of the lens which is numerically equal (but opposite in sign) to the focal length.

The focal length is related both to the material of which the lens is made and to the curvature of its surfaces. Let n be the refractive index of the lens material, and r_1 and r_2 respectively be the radii of the surfaces. If the medium on both sides of a thin lens is air then

$$f = v = \left(\frac{r_1 \cdot r_2}{(r_1 + r_2)(n - 1)} \right)$$

The magnification of a thin lens is given by

$$M = \frac{h'}{h} = \frac{v}{u}$$

Note that in the case of a simple converging lens forming a real image, in our convention v/u, h' and M are arithmetically negative, indicating that the image is inverted relative to the object.

1.4 REAL AND VIRTUAL IMAGES

A *real image* is one that can be seen on a screen or captured directly on a photographic film, while a *virtual image* can only be seen with the aid of another lens or lens system. If a simple microscope (single lens) is held close to the eye, it is not possible to measure the actual height of the real image which is formed on the retina by the microscope acting together with the cornea and lens of the eye itself. The magnification is under these conditions defined as the visual angle of the (virtual) image as seen by the eye with the aid of the lens, divided by the visual angle of the object which would be seen by the unaided eye at a conventional distance Dv of 250 mm. This arbitrary distance used to be called the *least distance of distinct vision*, a term which is, alas, inappropriate for most microscopists over the age of 40.

The calculated magnification of a simple microscope depends to some extent on whether the lens is used correctly in such a way that the muscles of accommodation of the eye are fully relaxed and the image appears at to be at infinity, or the image appears at the distance Dv. In the first case the object is situated at the front focal plane of the lens, v equals infinity and $1/f = 1/u$. The virtual image then subtends an apparent angle at the eye of h/f and

$$M = \frac{h/f}{h/Dv} = \frac{Dv}{f}.$$

In the second case the object is at such a distance to the left of the front focal plane of the lens that the virtual image is formed at 250 mm, and

$$M = \frac{h/u}{h/Dv} = \frac{Dv}{u}$$

The difference between the two cases is not large. For example, if $f = 250$ mm the magnification with the virtual image at infinity is $250/25 = \times 10$, while if the virtual image is at 250 mm, from the basic lens formula $u = -22.7$ and $M = 250/22.7 = \times 11$.

1.5   THE POWER OF LENS COMBINATIONS

The power in dioptres of a lens equals the reciprocal of its focal length measured in metres. If two thin lenses of focal lengths f_1 and f_2 are separated by a distance d, their combined power $1/f$ is given by

$$\frac{1}{f} = \frac{1}{f_1} + \frac{1}{f_2} - \frac{d}{f_1 \cdot f_2},$$

which reduces to

$$\frac{1}{f} = \frac{1}{f_1} + \frac{1}{f_2}$$

if the lenses are in contact.

1.6   HOW A SIMPLE LENS WORKS

The following is a qualitative summary of some salient facts. Figure 1.2 illustrates several rules of geometrical optics applicable to ideal, thin, positive (convex) lenses:

(a) Parallel beams of light striking a lens are brought to a focus in its back focal plane, at a distance behind the lens equal to the focal length of the lens. Beams of light parallel to the optic axis meet (and cross) at the intersection of the back focal plane and the optic axis.

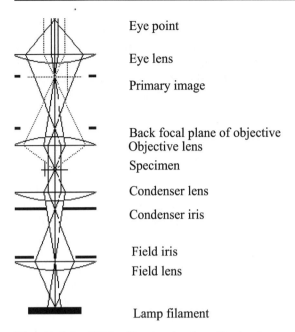

Eye point

Eye lens

Primary image

Back focal plane of objective
Objective lens

Specimen

Condenser lens

Condenser iris

Field iris

Field lens

Lamp filament

Figure 1.2   Köhler illumination (ray diagram)

(b) Rays radiating from a point on the front focal plane of a lens leave
the lens parallel to each other, and parallel to the optic axis if they
start off from the intersection of the front focal plane and the optic
axis. The distance of the front focal plane of the lens from the lens
numerically equals the focal length of the lens, but is conventionally
given a negative value.

(c) Rays pass undeviated through the centre of a thin lens, irrespective of
the angle at which they strike the lens.

Using these simple rules it is possible to draw a diagram showing
where an image of any object is formed. First, from a point on the object
draw a ray parallel to the optic axis, which on leaving the lens passes
through the intersection of the back focal plane and the optic axis. Draw
a second ray from the same object point to the intersection of the plane of
the lens and the optic axis, which then carries on undeviated. The image
of the point is situated where the two rays meet. If the object is further
away from the lens than the front focal plane, the image will be on the
opposite side of the lens, at a finite distance. If the object is exactly on the
front focal plane of the lens the two drawn rays leave the lens parallel to
each other and the image is at infinity; such rays can however be made to
converge to form a real image (i.e. one which can be seen on a screen or
captured on a photographic film) by adding a second convex lens to the
system. For example, if a microscope is used correctly, image-forming

rays leave the eyepiece parallel to each other, since the primary image is located on the front focal plane of the eyepiece; they are brought to a focus on the retina by the combined refracting action of the cornea and lens of the eye. If the object is closer to the lens than the front focal plane, a 'virtual' image will be present on the same (front) side of the lens, but further away than the object.

1.7   CONJUGATE PLANES

In an optical system, conjugate planes are imaged into each other and can therefore be seen simultaneously. It is important to distinguish between two sets of conjugate planes in the compound microscope, *field* (or *object*) planes on the one hand, and *aperture* planes on the other. The field stop adjacent to the lamp lens is imaged sharply into the plane of the object by the microscope condenser; images of both the object and the field stop are formed by the microscope objective in the 'primary image plane' in the front focal plane of the eyepiece, and the eyepiece (together with a microscope camera or the lens and cornea of the human eye) forms images of all three on photographic film or on the retina. The field stop, object, and primary image constitute the set of *object* or *image planes*, and when a light microscope is correctly adjusted a sharp image of the field stop ought to be clearly visible near the edge of the microscopic field. A disc of glass engraved with a linear scale ('graticule'), placed in the front focal plane of the eyepiece, is seen superimposed on the image of the object; such a graticule can be used as an *eyepiece micrometer* to measure dimensions in the object, and can be calibrated in absolute units with the aid of a suitably engraved *stage micrometer* situated in the conjugate plane where the object is normally found.

The set of conjugate *aperture planes* consists of the light source (e.g. lamp filament), the front focal plane of the condenser (where the condenser iris diaphragm is situated), the back focal plane of the objective, and the plane of the eyepoint in the back focal plane of the eyepiece.

Examining the back focal plane of the objective is essential in bright-field microscopy when centring the condenser and adjusting the diameter of the condenser iris diaphragm, and in phase-contrast microscopy when centring the annular condenser diaphragm relative to the phase plate of the objective. In practice, the back focal plane of the objective can be seen by taking out the eyepiece and (optionally) replacing it by a small, focussable telescope. In some instruments a built-in 'Bertrand lens' can be swung into place in the microscope tube, which together with the eyepiece (which is left in place) effectively forms a telescope. The back of the objective can also be visualized without taking out the eyepiece, by using a magnifying glass to examine the eyepoint just above the eyepiece.

The two sets of conjugate planes are essentially independent of each other, if the microscope is correctly adjusted. Closing the field stop therefore reduces the area of object illuminated but has no effect on the angle of the illuminating cone of light from the condenser. Conversely, closing the condenser diaphragm reduces the angular aperture of the light but not the area of specimen illuminated. Confirm these facts using the computer model, and if possible a real microscope.

1.8 FUNCTIONS OF THE FIELD AND APERTURE DIAPHRAGMS

Due to scattering of light, mainly by reflection at glass–air surfaces in the microscope objective, an appreciable fraction of the light from the object plane is often distributed diffusely over the image and results in a hazy image of low contrast. This stray light or glare can be reduced by careful cleaning of any accessible optical surface, and by limiting as far as possible the area of specimen which is illuminated. This is the function of the field diaphragm.

The diameter of the condenser iris diaphragm affects the intensity of the illumination striking the object but should not be used for this purpose; intensity should be controlled in other ways, for example by altering the voltage supplying the lamp or by inserting neutral filters in the optical path. The condenser diaphragm is used to control the angle of the cone of light striking the object, which has important and subtle effects on the imaging process, e.g. it influences both the resolution and the contrast of the system, often in opposite directions. This is not readily explicable by geometrical optics and is discussed in terms of diffraction theory elsewhere in this book.

1.9 KöHLER VS. 'CRITICAL' ILLUMINATION

The Köhler system of illumination (Figure 1.1), used in most modern microscopes, gives even illumination of a microscope specimen even if the lamp filament is not uniform. It can do this because it is not the irregular lamp filament which is imaged into the plane of the object, but the field diaphragm and the immediate adjacent lamp lens, which *is* uniformly illuminated.

Critical illumination (Figure 1.3, in which the eyepiece is omitted) is an alternative, somewhat old-fashioned system in which as before a field diaphragm is imaged into the plane of the object by the microscope condenser, but there is no lamp lens; an image of the lamp filament is therefore seen overlying the specimen. The area illuminated and the

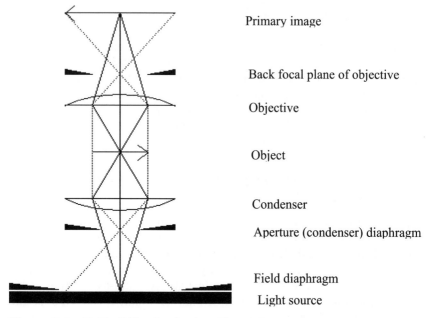

Primary image

Back focal plane of objective

Objective

Object

Condenser

Aperture (condenser) diaphragm

Field diaphragm

Light source

Figure 1.3   'Critical' illumination (ray diagram)

angle of the illuminating cone are controlled with field and aperture diaphragms just as with Köhler illumination. Although it is more difficult to obtain even illumination of the specimen using 'critical' illumination, Köhler and 'critical' systems are theoretically equivalent with respect to resolution, and the problem of uneven illumination of the object with 'critical' illumination can often be overcome by inserting a piece of ground, light-diffusing glass between the field diaphragm and the lamp filament. Placing the ground glass between the field diaphragm and the microscope condenser is theoretically incorrect, as this renders the field diaphragm ineffective. Even using Köhler illumination a piece of diffusing glass near the lamp filament may be useful, because otherwise an image of the filament is seen in the back focal plane of the objective – moderate non-uniform illumination of the aperture plane is theoretically undesirable but usually has little effect in practice.

1.10   EPI-ILLUMINATION (INCIDENT-LIGHT) MICROSCOPY

Transmitted-light microscopy is impractical and some form of epi-illumination must be used with very thick or dense objects such as bulk metals, minerals, ceramics and integrated circuits. 'Epi-illumination' refers to illumination which falls on the surface of the specimen nearest the objective lens, and although 'epi' really means 'from above' and can cause confusion in the case of an inverted microscope, 'epi-illumination' is marginally preferable to the alternative term 'incident-light illumination' since *all* illumination is incident on the object. I have fought a losing battle with colleagues to designate this type of illumination 'cis-illumination' in logical contrast with 'trans-illumination'.

It may be possible when using a low-power objective simply to shine a table-lamp or spot-light onto the specimen from the side, but with higher power lenses a specialized incident-light microscope is called for. Figure 11.1 is a simplified diagram of such an instrument. Note the field-stop, which limits the area of specimen illuminated and helps prevent stray light degrade the image. In practice a rather more complicated system is usually necessary, with provision for centring devices, an adjustable aperture stop and colour and/or polarizing filters.

In both incident-light and transmitted-light microscopy resolution is mainly determined by the wavelength of light and the numerical aperture of the objective (see Section 3.9), and the two methods have much else in common. Thus incident-light microscopy can be combined with numerous techniques discussed elsewhere in this book including fluorescence, phase-contrast, differential interference contrast, dark-field and polarized light.

Curiously, even though the specimen may look relatively familiar with incident-light illumination, light leaving the specimen can be affected by so many factors that interpretation of the image is often quite difficult. A detailed discussion must be left to those specializing in such matters, but a few general comments may be helpful.

Fluorescence or *phosphorescence* may be excited by the incident radiation, whether this is visible or ultra-violet. Reflection may be *specular* off smooth surfaces, *diffuse* off rough ones, or a combination of specular and diffuse. The *plane of polarization* of light can be affected by reflection or scattering, for example partially polarized light is produced if unpolarized light strikes a dielectric (non-conducting) surface at an angle, and elliptically polarized light may result if plane polarized light is multiply reflected and scattered off minute silver grains in autoradiographs (Goldstein and Williams, 1974).

The apparent colour of a specimen illuminated with white light depends in quite subtle ways on both the chemical nature and physical state of the material. *Surface colour* is seen with highly absorbing material such as metallic films, and consists of precisely those wavelengths which are most strongly absorbed by the material – reflecting power is closely related to absorption. *Body colour*, on the other hand, is produced if light is able to penetrate a little way into the specimen, so that those wavelengths which are most strongly absorbed are missing from the light which is eventually scattered or reflected back. For example, a dye which appears blue in solution does so because it has absorbed both red and green. If biological material is stained uniformly and moderately intensely with such a dye the reflected light is likely to show body colour, i.e. be blue. If the same dye is however present on the surface of the material in the form of crystals, the reflected light will probably show surface colour, i.e. be yellow (a mixture of red and green). It is also quite likely that light will be specularly reflected off such crystals, so that the azimuth of plane polarized incident light is effectively unaltered by reflection and a polarizing analyser can prevent light from the crystals reaching the image.

1.11   THE KOHLER PROGRAM

This is a non-quantitative, interactive computer program intended to aid the study of elementary geometrical optics, and how to adjust a bright-field or phase-contrast microscope correctly. The ideal lenses represented in the simulation are convex (converging) and 'thin' (free from aberrations).

1.11.1   Using the program.   Start the program in the usual way by typing RUN (ENTER), and pressing the letter K when you are asked to choose one of the four programs available. Pressing function key F8 while the program is running switches back to the menu offering a choice of programs. Any previous settings of the Zernike program are 'remembered' while in Kohler.

After reading the instructions and advice on the screen, and touching any key to proceed, a labelled ray diagram appears on the left of the screen showing the illumination system devised by Köhler (1894). This consists of the light source, lamp lens, field diaphragm just in front of the lamp lens, aperture diaphragm in the front focal plane of the microscope condenser, and the microscope condenser itself. The diagram includes a cross-shaped object, an objective lens with a fixed aperture stop in its back focal plane, the primary image in the front focal plane of the eyepiece, an eyepiece, and the eyepoint in the back focal plane of the eyepiece. Ray paths are drawn in different colours to illustrate some possible paths of light between the light source and the eyepoint. Light diffracted by the object is represented by dotted lines.

A diagram at the top of the screen shows the final image of the object and field stop or, if the eyepiece has been removed, the back focal plane of the objective with an image of the condenser iris (in phase contrast the condenser annular aperture and objective phase plate). On the right-hand side of the screen is a perspective diagram of a microscope, the controls of which can be operated either by pressing keys on the keyboard or by using the computer 'mouse' if one is present.

The controls available in the simulation are objective focussing, condenser focussing, condenser centration, adjustment of the diameter of the condenser diaphragm (in bright-field but not phase-contrast mode), and centration and diameter adjustment of the field diaphragm. These or similar controls are commonly found in real, good-quality microscopes, but some instruments may be slightly different – for example, it may be possible to centre the condenser but not the field stop. In phase-contrast mode, accessed by touching function key F3, the condenser aperture stop is annular and an annular phase plate is present in the back focal plane of the objective.

The various simulated microscope controls can be operated by the keys shown in Table 1-1:

Table 1-1

Component	Adjustment	Key
Objective	move up	A
	move down	B
Condenser	move up	C
	move down	D
Condenser (aperture) stop	open	E
	close	F
	move north-west	G
	move north-east	H
	move south-west	I
	move south-east	J
Fieldstop	open	K
	close	L
	move north-west	M
	move north-east	N
	move south-west	O
	move south-east	P
Eyepiece	remove/replace	Z

For example, on touching the letter A on the computer keyboard the objective lens is raised in the ray diagram, the ray paths change appropriately, the image is thrown slightly out of focus, and in the diagram of the microscope the microscope tube moves upwards. Touching the letter B restores the previous conditions. If the permitted limits of any movement are reached the computer 'beeps'. Operating the condenser and field-stop centring controls affects the final image and the microscope diagram, but only movements 'east' or 'west' are reflected by changes in the ray diagram since movements 'north' or 'south' are considered to be perpendicular to the plane of the VDU.

1.11.2   Use of the 'mouse'.   As an alternative to using the keyboard, with a suitably equipped computer the controls can be operated using the computer mouse. Place the mouse cursor (an arrow) on or near a red control knob or lever in the microscope diagram; the control turns brighter red, and the action of the control is described in a box on the screen (e.g. by UP, DOWN, OPEN, CLOSE, a clockwise arrow or an anticlockwise arrow). Pressing the left-hand mouse button operates the chosen microscope control, while pressing the right-hand button accesses the help function. The mouse cursor can be turned off at any time if not wanted, by pressing function key F7.

1.11.3   Saving screen graphics.   In all programs of the set, touching function key F9 saves the image on the graphics screen to a standard format .PCX file, the name of which you are asked to supply. This file can be subsequently loaded into various commercial graphics programs, modified if necessary, and printed.

1.12   EXERCISES USING THE PROGRAM

First study the initial, correct ray diagram and image, and investigate the effects of the various controls. Confirm the rules regarding the formation of an image by an ideal lens by changing the focus of the condenser or objective, and studying what happens to the rays as a result. Note which planes are conjugate with the specimen, and which with the back focal plane of the objective (if rays diverge from a point in one plane, and come together again in another plane, the two planes are conjugate).

1.12.1   Practice in setting up a microscope.   Randomise all adjustments by touching function key F5, and then attempt to restore the correct settings. Help can be obtained by touching function key F1, which causes the following suggestions to appear respectively for bright-field and phase-contrast microscopy. A green OK! or a red NOT OK! after each adjustment indicates whether or not the current adjustment is within acceptable limits.

Steps in setting up Köhler illumination:

(1) With the eyepiece in place, open the field iris if this is necessary to see the specimen, and focus the image of the specimen sharply with the objective focus.
(2) Partly close the field iris if this is necessary to see it, and focus its image with the condenser.
(3) Remove the eyepiece, and centre the condenser iris with the condenser controls.

(4) Adjust the condenser iris width to about 3/4 of the objective aperture.
(5) Replace the eyepiece and centre the field iris using the field iris centring controls.
(6) Open the field iris to show the whole object.

Steps in setting up phase contrast

(1) Unless the condenser is centred and focussed the transparent object will be invisible. Remove the eyepiece and focus the condenser.
(2) Centre the condenser.
(3) Replace the eyepiece, and open the field iris if this is necessary to see the object. Focus the object with the objective focussing controls.
(4) Partly close the field iris if this is necessary to see it, and focus its image with the condenser.
(5) Centre the field iris using the field iris controls.
(6) Open the field iris to show the whole object.

Chapter 2

Lenses and Lens Aberrations (the Snellius Program)

We have thus far implicitly assumed that image formation by a lens is effectively perfect, i.e. the image accurately resembles the object in every respect except perhaps size and orientation. In real life, however, even apart from the wave considerations which apply with small objects, an image formed by lenses with only spherical surfaces is invariably to some extent imperfect – the defects are called *aberrations*. It is impossible to eliminate all aberrations completely from a spherical lens system, and the art of the lens designer consists of reducing as far as possible, consistent with price, those defects which have the most harmful effect on the intended purpose of the lens.

In the design of lenses it used to be the practice to use mathematics (in general based on Gaussian optics) initially and then to test and refine the results by the use of ray-tracing; nowadays the use of computers in commercial lens design is universal. In our discussion of real lenses and their aberrations we shall skip the difficult algebra, and adopt a non-mathematical, qualitative approach plus the use of simple computer-aided ray-tracing.

Aberrations seldom or never occur in isolation, but can be broadly classified into those which affect on-axis as well as off-axis images, and those which only apply to off-axis ones. In the former group are (longitudinal) chromatic aberration and spherical aberration, while the latter group includes astigmatism, coma, distortion, curvature of the field, and chromatic difference of magnification.

2.1 CHROMATIC ABERRATION

The refractive index of a solid or liquid transparent medium varies with the wavelength of light, so that white light passing through a prism or simple lens is dispersed (split) into a coloured *spectrum*. The refractive index is generally higher with shorter wavelengths, i.e. light towards the blue end of the spectrum is refracted more than red light. The effect is that images formed with light of different colours are formed at different points along the optic axis, and at a given focus the image is surrounded by coloured fringes. This is *longitudinal chromatic aberration*, which should not be confused with chromatic difference of magnification (discussed later).

The chromatic aberration of even a relatively poor lens system can sometimes be reduced to acceptable limits in microscopy (e.g. in black-and-white photomicrography) by using a coloured filter to obtain effectively monochromatic light For most purposes it is however necessary to control chromatic aberration by appropriate lens design. Newton discovered prismatic dispersion in 1666, and on the basis of some rather unfortunate experiments he concluded that dispersion is directly proportional to refraction and that it is therefore impossible to correct chromatic aberration using lenses. He accordingly recommended the use of reflecting optics, which are not subject to chromatic effects. In fact, dispersion due to refraction varies with different types of glass, and it is possible to correct dispersion without completely eliminating the bending of the rays by a lens system. Although the relationship of refractive index to wavelength is not in general strictly linear, if linearity is assumed dispersion can be characterised approximately by the *dispersive power* Δ defined by

$$\Delta = \frac{n_F - n_C}{n_{D-1}}$$

where n_F, n_D and n_C are respectively the refractive indices for the Fraunhofer F, D and C lines (0.4861, 0.5893 and 0.6563 μm). For example, the typical dispersion of flint glass (ca. 1/30) is nearly twice that of ordinary crown glass (ca. 1/60). An *achromatic doublet* with a net converging power but very little chromatic aberration can therefore be made by combining a weak concave (negative) flint glass lens with a stronger convex (positive) crown glass lens. The mineral *fluorite* is particularly useful in the design of advanced microscope objectives because of its unusually low dispersion; it is unfortunately birefringent (see Chapter 14), which can affect the performance of lenses used in polarised-light work.

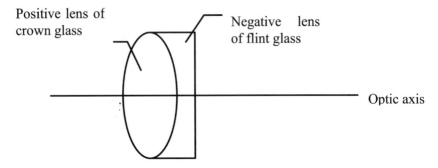

Figure 2.1   Achromatic doublet

2.2 SPHERICAL ABERRATION

A ray of light parallel to the axis of a lens with spherical surfaces is in general refracted more strongly the further it lies from the axis (see Figure 2.2). The result is that a ray striking near the edge of a lens crosses the axis nearer the lens than does a paraxial ray. This *longitudinal* aberration should not to be confused with *curvature of the field*, discussed later. An *aplanatic* lens is one which is effectively free from both spherical aberration and a related defect called *coma*.

For a given strength of lens the spherical aberration depends to a large extent on the way the curvature is distributed between the surfaces. Thus a biconvex lens, situated near an object and giving an enlarged distant image, exhibits considerably more spherical aberration than a plano-convex lens of similar strength in the same position, oriented with its flat side nearer the object. Conversely, with a distant object and a relatively near image, a convex-plano lens with its flat side facing the image produces an image with less spherical aberration than do plano-convex or biconvex lenses of similar strength.

Although it is possible to construct a simple lens which exhibits minimum spherical aberration for an object at a particular distance, no single spherical lens can be completely free from spherical aberration. An aplanatic combination of two such lenses *can* however be constructed. An

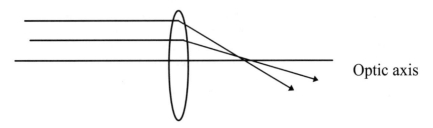

Figure 2.2   Spherical aberration

example of such a combination is a relatively strong positive lens which gives minimum spherical aberration for an object at a particular distance, combined with a weaker negative lens made of the same type of glass, but constructed to give maximum (negative) spherical aberration. Together the two lenses have a net converging action but much reduced or no spherical aberration, at least for light passing through selected zones of the lens. Thus light passing very close to the optic axis (paraxial light) can be brought to the same focus as light passing through the lens near to its edge; although light passing through intermediate zones may not come to exactly the same focus, this residual spherical error may be small enough to be acceptable.

If the two lenses of an aplanatic doublet are cemented together the combined lens is stable and durable, and (as pointed out by Lister in 1830) reflections at the glass-air surfaces are considerably reduced, but the lens designer then only has three radii to play with since lens surfaces which are cemented to each other must have the same radius of curvature. If the two lenses can be separated by a small distance the designer has more freedom, since he or she has four independent radii instead of three, and in addition can vary the distance apart of the two lenses, which critically affects the magnitude of the spherical correction. This fact is exploited in high-powered 'dry' lenses which are fitted with a *correction collar*; this enables the separation of the elements to be controlled by the user, who can adjust the lens to give minimum spherical aberration with a coverglass of slightly incorrect thickness.

If the two lenses of an aplanatic pair are made of different types of glass it is possible to correct for chromatic aberration at the same time as spherical aberration.

An aplanatic combination of lenses in general has two pairs of *aplanatic points*. An object at one point of such a pair gives an image, more or less free from spherical aberration, at the other point of the pair. One of the pairs of aplanatic points gives a real image on the side of the lens away from the object, while the other pair of aplanatic points gives a virtual image on the same side of the lens as the object but further from the lens. In the early 19th century Lister (who was father of the surgeon who became famous for promoting antiseptic surgery) introduced into microscopy an objective consisting of two achromatic doublets, constructed in such a way that it was relatively free from both chromatic and spherical aberration. The object was situated close to the objective in a position so that the first achromatic doublet formed an enlarged virtual image, on the same side as the object, which served as the object for the second achromatic doublet. This formed a real image in the front focal plane of the eyepiece. Similar constructions are still used in modern low-power (say × 10) objectives intended for student use – in some cases the front lens of the pair can be unscrewed, leaving the upper lens to act alone as an objective of even lower power.

The main problem with this type of objective is that its angular aperture (the maximum angle of light which it can receive from the object) is severely limited. We shall see later that this is a crucial factor in determining the resolving power and light-gathering capability of a lens.

Amici (1840) further improved objective corrections by using the following principle, probably due originally to Huygens.

Consider a glass sphere of refractive index n and radius r (Figure 2.3). Light emerging at any angle from a point situated inside the sphere, at a distance r/n from the centre, will have a virtual image, completely free from spherical aberration, at a distance rn from the centre. The proof of this involves only elementary trigonometry. In practice it is not feasible to embed an object inside a glass sphere, but the same optical effect may be obtained by truncating the sphere and mounting the object in immersion oil of the same refractive index as the glass of the lens, in contact with the front of the lens. Alternatively, the sphere can be replaced by a meniscus, the front surface of which has a radius of curvature with the object at the centre – light from the object then passes into the glass undeviated, and is refracted aplanatically by the second surface of the lens. An intrinsic limitation of this type of lens is that it can only form a virtual image (i.e. it cannot by itself converge the light sufficiently to form a real image). The virtual image can however be further enlarged by a second aplanatic meniscus, and finally a real image is formed by one or more achromatic doublets of the type already described.

It should also be noted that the Amici front element is aplanatic only for light of a particular wavelength, and chromatic correction must be dealt with by the other lenses of the system.

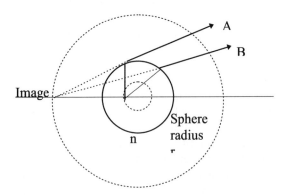

Figure 2.3   Principle of Amici's aplanatic lens

The maximum half-angle of light from the object (measured relative to the optic axis) which can be trapped by an Amici lens is 90°, corresponding to a numerical aperture of 1.0 for a dry lens (see line A in the diagram). At any angle greater than 90° the light from the object is totally internally reflected at the glass–air interface and cannot take part in image formation. Even an acceptance angle of 90° however implies the uncovered object is in direct contact with the front of the lens, which is not practical for ordinary use. The maximum practical value for the numerical aperture of a dry lens is about 0.95, corresponding to an acceptance half-angle of just over 70°.

Objective lenses are corrected for objects and images at particular distances, and their spherical correction can be affected by a variety of factors. Oil-immersion lenses and low-power dry lenses are relatively insensitive to cover-glass thickness, but high-power dry lenses intended for use with covered objects are usually designed for cover-glasses about 0.17 mm thick, and their spherical aberration can be markedly increased by use with an uncovered object, or a coverslip of an incorrect thickness. An incorrect tube-length is another possible source of spherical aberration, which can either aggravate or compensate error due to the cover-glass. Adjustment of the tube length is seldom possible in modern microscopes, but as mentioned above some high-power dry objectives are equipped with a correction collar which enables the spherical correction of the lens to be altered by changing the distance apart of lens components.

Spherical aberration can in practice be recognised with the *star test*. A small opaque object in a bright field, or a pin-hole aperture in an opaque metal film, is carefully focussed in the middle of the microscopic field, and the microscope focussed alternately up and down. In the absence of spherical aberration the image enlarges symmetrically on both sides of the best focus, while in the presence of spherical aberration the (blurred) image of the point is larger on one side of the focus than on the other. An experienced microscopist can use the star test to adjust the correction collar of his or her high-powered objective. Cynical (or disillusioned) microscope manufacturers nowadays unfortunately often omit correction

Table 2.1 Spherical aberration

Cause	Undercorrection	Overcorrection
Coverglass	Too thin	Too thick
Tube	Too short	Too long
Correction collar	Lens components too close together	Components too far apart
Immersion oil	Refractive index too low	Refractive index too high
Diagnosis	Greater expansion of image on focussing down	Greater expansion on focussing up

collars from lenses intended for routine use, since a lens with an incorrectly adjusted correction collar gives substantially worse results than a cheaper lens lacking such a feature.

2.3   COMA

Coma (which literally means *hair*) is an aberration closely related to spherical aberration. It is due to the fact that the refraction of light from an off-axis object differs for different zones of a spherical lens. In the presence of coma the image of an off-axis point-like object appears as a number of overlapping non-concentric circles with radii which increase with distance from the optic axis. The result is a comet-shaped flair which in general lies with its wide, blurred tail on the side away from the optic axis. The size of the blurred image increases approximately proportionately to the distance of the object from the optic axis. Coma is a serious error, and in the design of lenses much attention is paid to its elimination. It is therefore seldom or never found in good-quality lenses.

2.4   ASTIGMATISM

A spherical lens surface tends to act more strongly on a ray inclined at a large angle to the optic axis, than on one parallel to the axis. The effect of this is that the image of an off-axis point object, instead of being a point (or Airy disc if one takes the wave nature of light into consideration) may be a line, the direction of which depends on the focus of the lens. At one level of focus the line points towards the optic axis, while at another focus the line is at right angles to the first (Figure 2.4). At some intermediate level (the position of *best focus*) the image will be approximately round, and of minimum size.

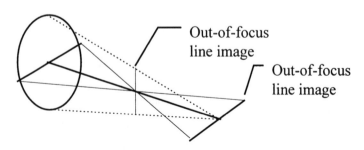

Out-of-focus line image

Out-of-focus line image

Figure 2.4   Astigmatism of off-axis image

2.5 CURVATURE OF THE FIELD

In this aberration, often found in association with astigmatism, the image of a thin plane object falls not on a plane but on a curved surface.

Object perpendicular to
optic axis

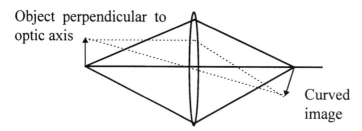

Curved
image

Figure 2.5 Curvature of the field

If an object in the middle of the field is in focus, an off-axis object will therefore be out of focus and *vice versa*. Significant curvature of the field is extremely troublesome with photomicrography, but a moderate amount may be tolerable with visual observation since all parts of the field can be seen sharply in succession simply by using the fine adjustment of the microscope.

2.6 DISTORTION

This aberration can be caused by incorrect placement of stops in a complicated lens system, which is usually outside the control of the user. It should be noted that even if there is no obvious metal stop in the system, the edges of any lens are potential stops. Distortion is seldom significant in a single lens without a separate stop, but can occur in a combination of two such lenses.

In the presence of distortion any straight line in the object which crosses the optic axis is imaged as a straight line, but other straight lines in the object appear curved in the image. Two forms can be recognised (Figure 2.6). In *barrel* distortion the image of a uniform grid-like object is deformed in such a way that parallel lines of the object appear in the image concave towards the optic axis, while in *pin-cushion* distortion parallel lines in the object appear in the image convex towards the optic axis. Distortion can seriously affect quantitative microscopic work including stereological measurements, but is of less importance for simple visual observation.

To understand how distortion might come about (Longhurst, 1973, p. 394), consider a lens which produces an image suffering from appreciable curvature of the field (Figure 2.7). With a stop placed next to the lens, or with no stop other than the lens mounting, rays from an off-axis object point will cross some distance away from the plane where the centre of

Barrel   Pin-cushion

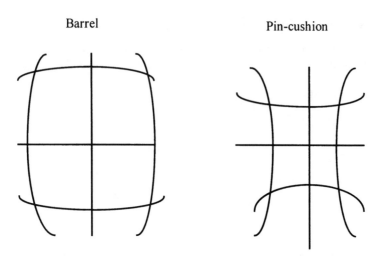

Figure 2.6   Barrel and pin-cushion distortion

the object is in focus, and produce a rather fuzzy blur in that plane with the centre of the blur at point a. Now imagine that a stop is placed in between the lens and the object, on the optic axis, of such a diameter that it allows through only a small part of the cone of light from the object point in question. The image of the object point is now sharper, since it is formed only by a few rays, but it will be formed at point b, which is nearer the optic axis. The result is pin-cushion distortion. Similarly, a stop on the other side of the lens (near the image plane) would allow through only ray c, and cause barrel distortion.

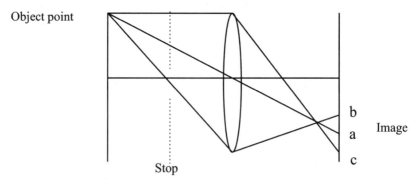

Figure 2.7   Origin of distortion

2.7   CHROMATIC DIFFERENCE OF MAGNIFICATION

As the name implies, images formed by light of different colours are of different sizes, resulting in coloured fringes at the edges of images particularly towards the edges of the field. The aberration is sometimes dealt with in microscopy by the use of *compensating eyepieces* (see below).

2.8 TYPES OF OBJECTIVE

2.8.1 Achromats, semi-apochromats and apochromats.

The terminology of objectives is to some extent affected by commercial usage, but in general *achromatic objectives* (achromats), which are relatively inexpensive, are corrected for spherical aberration for one colour (usually green, to which the eye is most sensitive), and are corrected for chromatic aberration for two colours. *Apochromats*, designed originally by Abbe around 1886, are relatively complicated and hence more expensive than achromats, and use fluorite and several different kinds of glass. Fluorite is a natural mineral with the valuable property of having a very low dispersion. Spherical aberration is corrected in apochromats for two wavelengths, and chromatic aberration for three wavelengths. Other corrections are also usually better in apochromats than in achromats, and the numerical aperture of an apochromat of a given magnification is also usually larger than that of an equivalent achromat.

Abbe found it impossible to eliminate chromatic difference of magnification in high-power apochromatic objectives, and introduced special compensating eyepieces for use with such objectives to correct this aberration. In order to be able to use the same eyepieces with all apochromats, Abbe deliberately introduced chromatic difference of magnification into his lower-power apochromats. Compensating eyepieces can be distinguished from ordinary (e.g. Huygens) eyepieces by the colour of the fringe visible at the edge of the field stop – this is orange in the former and blue in the latter.

Used with a green filter the performance of a good achromat may be as good as that of a much more expensive apochromat, and is sometimes even better as achromatic lenses are simpler and therefore less liable to glare (stray light), which degrades contrast in the image, and to strain birefringence, which can affect observations using polarised light. Apochromats are however usually preferable for use with white light, or light at the extremes of the spectrum.

Fluorite lenses, like apochromats, include a fluorite component. They are however of simpler construction than apochromats and are intermediate in quality and price between achromats and apochromats. Plan (flat-field) objectives, available in achromat, fluorite and apochromat versions, are especially valuable for photomicrography. The best modern objectives have corrections which are considerably better than those of older lenses – this is a benefit of the computer-aided lens design systems now in general use.

2.8.2 Infinity-corrected objectives. In many modern microscopes the object is situated precisely in the front focal plane of an objective which is designed to give a well-corrected primary image at infinity. The concept of optical tube length therefore no longer applies. Rays leaving the objective parallel to the optic axis are brought by a *tube lens* to a focus in the front focal plane of the eyepiece. This system conveniently provides more room in the microscope tube for the insertion of such accessories as polarisation compensators or analysers.

2.9 THE SNELLIUS PROGRAM

This is a simple ray-tracing computer program in which the paths of rays through a variety of lenses can be studied. The lenses provided include converging (positive) biconvex, plano-convex, convex-plano and doublet types, a diverging (negative) lens and an aplanatic lens of the Amici type. The user is also able to define a set of up to five lenses of arbitrary properties (curvature, width, height, position, refractive index and dispersion). User-defined lenses can be saved to disc for later use.

Rays start off from one of several objects. These include an axial object, a distant object providing rays parallel to the optic axis, an off-axis object giving both meridional rays in the same vertical plane as the optic axis and skew rays at an angle to the optic axis, and an object giving rays both from its base on the optic axis and from its head above the axis. User-defined objects can be generated, and if desired saved to disc. The paths of meridional rays through the chosen lens or lenses are shown on the computer screen, and optionally a 'face-on' view is also shown of a 'screen' perpendicular to the optic axis, on which the points of impact of both meridional and skew rays can be seen.

The position of the lens or lenses and/or the screen relative to the source of the rays can be adjusted (focussed). Either monochromatic green or white light can be specified, the latter actually consisting of a mixture of blue, green and red beams, each of which is subject to a different refractive index in the lens or lenses. Optionally the final rays leaving the lens system can be traced backwards to demonstrate the position of a virtual image, if present. Total internal reflection at surfaces is recognised and indicated by the program, but partial reflections are ignored.

A Help function provides information on the operation of the program, Snel's law and various aberrations. Exercises are suggested to study many of the aberrations discussed.

Chapter 3

Elementary Diffraction Theory (mainly due to Abbe)

Early in the 19th century pioneers including Fraunhofer and Airy attempted to analyse microscopic imaging in terms of diffraction and interference. The most important contribution to the theory of the microscope was however made by Ernst Abbe (1873, 1904). Abbe was unfortunately not a prolific writer, and the first systematic account of his work was given by Lummer and Reiche (1910; see also Michel, 1964; Martin, 1966). An independent and apparently different theoretical approach taken by Rayleigh (1896) can be shown (Born and Wolf, 1975, p. 420) to be mathematically equivalent to that of Abbe.

The following elementary explanation of the microscopic imaging of non-self-luminous objects, illuminated with coherent light, is based on the work of Abbe. The minimal mathematics involved may be ignored without serious loss by those averse to such things. The theory is later extended to incoherent and partially coherent illumination, and to the imaging of self-luminous (e.g. fluorescent) objects.

3.1 INTERFERENCE AND THE WAVE NATURE OF LIGHT

Under some conditions light behaves as if it consists of separate particles, corpuscles or *quanta*. This is true, for example, when light is produced in a lamp, or converted into other forms of energy such as heat in a light-absorbing filter or chemical energy in a photographic film. In dealing with microscopic image formation, however, light is best regarded as being energy in the form of transverse waves. These waves are characterised by wavelength (or frequency), speed, amplitude (or intensity), phase, and plane of polarisation. The *frequency* of a given wave is measured in cycles per second and is invariant, but *wavelength* (the distance from one peak of a wave to the next) and *speed* (wavelength multiplied by frequency) are inversely proportional to the refractive index of the medium in which the light is travelling. Wavelength and frequency are related to the perceived colour. The *amplitude* of a wave is the height of its peaks above, or the depths of its troughs below, a baseline; the sign of the amplitude is positive for peaks and negative for troughs. *Intensity* equals the square of the amplitude, and is always positive. *Phase* refers to the position of the peaks (or troughs) of a wave relative to those of a *reference* wave, and can be expressed in degrees, radians or fractions of a wavelength. Thus a *retardation* of a quarter wavelength (90° or $\pi/2$ radians) means that the light has been delayed relative to the reference wave by one quarter of the distance between two adjacent peaks; this is in general equivalent to a relative *advance* of three-quarters of a wavelength. A full wavelength (360° or 2π radians) retardation brings a light wave back into perfect alignment with the reference beam, and is for many purposes equivalent to zero retardation. *Polarisation* is the direction of the transverse vibration of the wave when viewed 'end-on'; the existence of polarisation is the best evidence for the transverse nature of light waves, as opposed to the familiar longitudinal waves found in water in which at least some of the movement of the water molecules is in the direction of movement of the energy. Wave packets ('photons') given off successively by a point source such as a small part of an incandescent tungsten filament, or simultaneously by independent point sources, vary unpredictably in wavelength, phase and polarisation and are described as *incoherent*. To be capable of *interfering* visibly to give a single *resultant* wave, waves must be *coherent*. That is, they must have the same wavelength as each other, and maintain a constant phase and polarisation relationship with each other over a fairly long period. Coherent beams can be produced by dividing a beam of light into two with the aid of a beam-splitter such as a half-silvered mirror. The *relative* wavelength, phase and polarisation of the two beams then remain constant even if the properties of different wave-packets vary, and if the beams are re-united (say by a second half-silvered mirror) interference may be seen. Coherent light is also produced by a laser, but this is outside the scope of the present discussion.

Lines joining points of equal phase in a beam of coherent light represent *wavefronts* which lie at right angles to the direction of movement of the energy. If two coherent waves in phase with each other are brought together they interfere *constructively* to give a resultant, the amplitude of which equals the sum of the original amplitudes. Conversely, if two interfering waves are perfectly out of phase (have a phase difference of half a wavelength) their interference is mutually *destructive*. In intermediate cases the resultant can be calculated by methods described below (see Section 3.8). Conservation of energy ensures that light 'missing' from one place due to destructive interference re-appears due to constructive interference somewhere else, often not very far away.

3.2 DIFFRACTION BY AN AMPLITUDE GRATING

Consider a parallel beam of coherent light which is normally (i.e. perpendicularly) incident on an extended, uniform grating consisting of narrow, transparent slits and opaque bars. of equal width (Figure 3.1). Such a grating is sometimes called an *amplitude grating* because it affects the amplitude but not the phase of the light passing through it. The *interval* of the grating, i.e. the distance between the centres of adjacent slits, can be measured in micrometres or more conveniently for our purpose in fractions or multiples of a wavelength (λ). Light reaches all parts of the grating in phase, and according to Huygens' principle (Born and Wolf, 1975) each slit may be considered to emit concentric, spherical

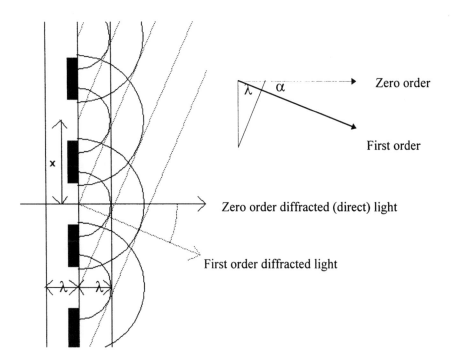

Figure 3.1 Diffraction at an amplitude grating

wavelets of light which start off in phase with each other. Wavefronts may, as before, be represented by lines joining points of equal phase; in this case the wavefronts are tangents to adjacent spherical wavelets. Wavefronts leaving the grating parallel to the wavefronts of the original illumination constitute the *direct* or *zero-order diffracted* light. It is also often possible to draw wavefronts in such a way that wavelets from a given slit are shown to be in phase with wavelets from an adjacent slit, one whole wavelength behind or in front – remember that a whole wave phase difference means that the beams are actually in phase with each other. This new wavefront, *first-order diffracted light*, is inclined at an angle $\alpha = \arcsin(\lambda/x)$ relative to the direct light, where λ is the wavelength and x is the interval of the grating. If x is fairly large relative to λ, additional (second or higher) orders of diffracted light may also occur, inclined at yet greater angles. For clarity, Figure 3.1 shows only the direct light and the first-order diffracted light on one side of the optic axis, but diffracted beams should actually be symmetrically present on both sides of the axis.

3.3   DIFFRACTION BY A 'NON-EXISTENT' GRATING

Imagine a *virtual* (actually non-existent!) grating in which both the bars and the slits are perfectly transparent, and have the same refractive index (Figure 3.2). The point of this apparently futile exercise should become evident very shortly.

Direct and first-order diffracted beams from this hypothetical grating are theoretically generated by the light from the 'slits' just as in the case

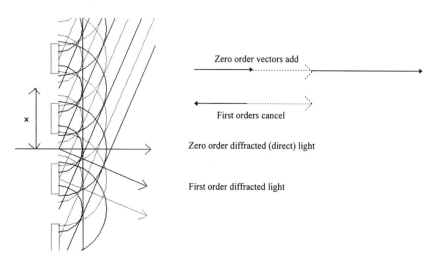

Zero order vectors add

First orders cancel

Zero order diffracted (direct) light

First order diffracted light

Figure 3.2   Diffraction at a virtual grating

of an amplitude grating, and exactly the same applies to the 'bars'. The zero-order diffracted light beams from the slits and bars are in phase, and interfere constructively to give a single direct resultant. The first-order (and higher-order) diffracted beams from the bars are however precisely half a wavelength out of phase with those from the slits, with which they therefore interfere destructively, leaving only the direct light, as common sense has already predicted!

This explanation of how light waves travel in straight lines through empty space is adequate for our present purposes, but has been somewhat simplified. It ignores a quarter-wave phase difference between the secondary wavelets and the original wave, and the fact that the amplitude of the wavelets is a function of their inclination to the direction of travel of the energy (see Born and Wolf, 1975, pp. 370–380).

3.4 ABBE'S THEORY OF MICROSCOPIC IMAGE FORMATION

We now have just about enough understanding of diffraction to be able to tackle Abbe's theory. According to Abbe, image formation in the compound microscope takes place in two distinct stages. In the first of these, light incident on an object is diffracted into discrete beams (orders). The zero-order light reaches the objective travelling parallel to the optic axis, and is therefore brought to a focus in the centre of the back focal plane of the objective. The first-order beams of diffracted light, inclined relative to the optic axis, are focussed by the objective into spots on either side of the optic axis in the back focal plane of the objective. A pattern of light spots is therefore visible in the back focal plane of the objective, and it can be shown that the distance from any given point in this pattern to different points in the object plane is linearly related to the position of the object point. The pattern is therefore by definition a *Fraunhofer diffraction pattern* (Longhurst, 1973 p. 226), but was called by Abbe the 'primary interference image' (Bradbury *et al.*, 1989) and is often referred to in modern texts as the Fourier transform of the real image. We may note in passing that at other distances from the object, measured along the optic axis, the diffraction process is referred to as Fresnel diffraction – the Fraunhofer diffraction pattern is merely a special case (see Section 3.11).

The physical existence of the Fraunhofer diffraction pattern can readily be confirmed using a low-power microscope objective, a suitable grating object (e.g. a commercial stage micrometer or a piece of fine wire mesh), and a very small condenser aperture. If you remove the eyepiece and look down the microscope tube, the direct light can be seen as a bright central spot in the back of the objective, flanked on either side by slightly less bright spots (the first-order beams). Additional spots of higher order are present further from the optic axis. The spacing of the spots depends on the interval of the grating and the wavelength of the light used – the smaller the interval of the grating or the longer the wavelength of the light, the further apart are the spots. If the light is white (a mixture of many different wavelengths) the longer-wavelength components are diffracted through a greater angle, so that each lateral spot is actually a spectrum with blue on its axial and red on its outer side.

In the second stage of the imaging process, coherent light beams emitted by different parts of the diffraction pattern mutually interfere to produce a real 'primary image' situated in the front focal plane of the microscope eyepiece. The eyepiece has the relatively uninteresting role, which we will not discuss here, of enlarging and inverting the primary image before detection by the eye or camera.

The primary image is in a sense produced not by the object itself but by the Fraunhofer diffraction pattern. The extent to which the image resembles the object therefore depends, among other factors, on how much of the light diffracted by the object reaches the diffraction pattern, and the image can be radically affected by using an objective of small aperture which excludes some diffracted light.

Abbe pointed out that for *any* recognisable image to be formed by interference the objective must capture not only the zero-order light but also at least one of the first-order beams – the zero-order beam would by itself merely illuminate the image plane uniformly. The angular aperture of the objective must therefore be at least large enough to admit the zero-order and the first-order beams. Since $\alpha = \arcsin(\lambda/x)$ (Figure 3.1), the finest grating which can be resolved under the defined conditions is

$$x = \frac{\lambda_n}{\sin(\alpha)} \tag{3.1}$$

Here λ_n is the wavelength of light measured in the medium, of refractive index n, which lies between the object and the objective. Alternatively,

$$x = \frac{\lambda_0}{n.\sin(\alpha)} \tag{3.2}$$

where λ_0 is the wavelength measured in vacuum. Abbe defined the *numerical aperture* (NA) of the objective as

$$NA = n.\sin(\alpha) \tag{3.3}$$

so that

$$x = \frac{\lambda_0}{NA} \tag{3.4}$$

This equation, which is of fundamental importance in microscopy, shows that the larger the objective NA or the smaller the wavelength,[1] the finer is the grating which can be resolved. The reason why oil-immersion and water-immersion objectives can have numerical apertures greater than unity, and hence give better resolution than any 'dry' lens, is also apparent – the angle of the cone of light which immersion lenses are able to capture is no greater than that of a good dry lens, but the wavelength of light *in the immersion medium* is shorter. Note that what is important is the lowest refractive index of any medium between the specimen and the front element of the objective,[2] so that if an object is embedded in (say) a plastic with a high refractive index but there is a layer of air between the coverglass and the objective, the refractive index of air limits the numerical aperture and the resolving power.

A grating with an interval of precisely one wavelength, illuminated with coherent light, would just be resolved by a (dry) objective with a numerical aperture of 1.0 (corresponding to a theoretical angular half-aperture of 90°). The maximum possible NA of a real (dry) objective is about 0.95, so that with axial coherent light no image can ever be formed of a grating having an interval less than one wavelength. In the computer program the unit of distance in the object plane is a half-wavelength, which permits a grating to be simulated which has a one-wavelength interval with minimum-width slits separated by minimum-width bars.

Let us look more closely at image formation by the zero-order and first-order beams. The *optical path* travelled by light is defined as the geometrical distance covered, multiplied by the refractive index of the medium. Elementary optical theory teaches us that light from a given point in an in-focus object travels through exactly the same total optical path to reach the image plane, irrespective of the route taken. Light passing through a converging lens near the optic axis of course travels through a shorter geometrical distance than that which passes near the edge of the lens, but the difference is exactly compensated for by the fact that light near the axis travels further through the substance of the lens, which has a relatively high refractive index.

All the light leaving the amplitude grating starts off in phase. Assuming that a slit of an amplitude grating is located on the optic axis, the optical distance from the central grating slit to the centre of the Fraunhofer diffraction pattern is identical with that to the first-order spot in the diffraction pattern; the distances from these spots to the centre of the image plane are also identical with each other. When they reach the

---

[1] Hamlet talked of the '... native hue of resolution ...'. He appears to have meant a rather ruddy colour, but in microscopy (as opposed to Elsinore) resolution is best at the blue end of the spectrum.
[2] A widely believed anecdote concerns a medical student who when using an oil-immersion lens for the first time asked the demonstrator for a second bottle of oil, because the one he had been given did not contain enough to fill the microscope tube completely between the objective and the eyepiece.

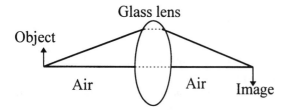

Figure 3.3   Optical paths via a lens

centre of the image plane the zero- and first-order diffracted beams are therefore still in phase, so that they interfere constructively to give a bright line. Elsewhere in the image plane the situation is different. For example, at some lateral point in the image plane the distance to the centre of the Fraunhofer diffraction pattern is precisely half a wavelength longer or shorter than the distance to one of the first-order spots. At such a lateral point, therefore, the zero- and first-order beams interfere *destructively*. Yet further laterally in the image plane the arriving zero- and first-order beams differ in phase by a full wavelength, and the interference is again constructive. At intermediate image points the interference is neither completely constructive nor completely destructive. It can be shown that the image consists of a series of alternating bright and dark bands, with the same spacing (allowing for the magnification of the system) as in the original grating. The image is a (possibly crude) representation of the object. If an image is formed by only the zero- and first-order diffracted beams, the intensities in the image vary sinusoidally, and the change from dark to light is not abrupt as we postulated in the object. If higher orders of diffracted light are allowed to take part in the formation of the image, the image comes to resembles the object more and more closely.

A little consideration shows that if instead of a slit, an opaque grating bar is situated on the optic axis in the object plane, the direct and first-order beams in the diffraction pattern must be half a wavelength out of phase with each other. The image consists as before of bright and dark lines or bands, but this time with a dark line on the optic axis.

3.5   EFFECT OF OBLIQUE COHERENT ILLUMINATION ON THE RESOLUTION OF A GRATING

Image formation normally requires the participation of the direct light and at least the first-order diffracted light. Because the two first-order beams (one on either side of the optic axis) carry identical information, an image can however be formed by the direct light together with a *single* first-order beam. If the parallel beam of coherent illuminating light is inclined to the optic axis at such an angle that the direct light leaving the

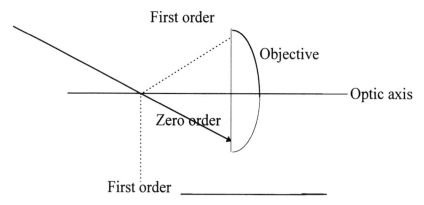

Figure 3.4   Resolution of grating with oblique illumination

specimen only just enters the microscope objective (Figure 3.4), the objective is able to capture a first-order beam diffracted at a greater angle than previously, even though the first-order beam on the opposite side is lost.

Under these conditions the maximum resolving power for a grating object is precisely doubled, and the resolution equation becomes

$$X = \frac{\lambda_0}{2NA} \tag{3.5}$$

3.6 EXTENSION OF THE ABBE THEORY TO NON-PERIODIC OBJECTS

Gratings are useful test objects in microscopy, but few natural biological specimens other than diatoms and striated muscle exhibit marked periodic structure. To what extent does Abbe's theory apply to ordinary objects? A traditional answer is that it is possible to analyse any image, no matter how complicated, into overlapping sinusoidal gratings of varying intervals and amplitudes. Strictly speaking, according to Fourier's theorem any repeated spatial function with a period x can be analysed into (or synthesised from) a set of harmonic functions with wavelengths of x, x/2, x/3, etc. The mathematical procedure for doing this is the Fourier transform – see Section 15.2. Conclusions derived from studying a simple grating ought then to be in general valid for individual components of the image of any specimen, and presumably also for the whole image. Such a broad generalisation may however be going rather too far – for example, we shall see that the limit of resolution for two adjacent points or lines is *not* necessarily the same as for a grating.

The following informal discussion of the imaging of non-periodic objects is based loosely on the approach of Rayleigh (1896, summarised in Martin, 1966, pp. 272 *et seq.*). We start by discussing the image of a single point (Exercise E3).

3.7 THE AIRY DISC

From a single illuminated point such as a minute pin-hole in an opaque silver film, light spreads forwards with a spherical wavefront. If such a point is in focus on the optic axis of a microscope, after passing through the objective lens the spherical wavefront becomes convergent with its centre of curvature in the middle of the image plane, where it interferes constructively. If *all* the light from the object were captured by the objective (which is actually physically impossible, since it would require an acceptance angle of 180°, the image would consist of a bright point on the optic axis surrounded by a dark background. Why is the background dark? This is due to the fact that in the image plane only the optic axis is equidistant from every part of the curved, evenly illuminated Fraunhofer diffraction pattern, so that only here is the interference of light from the pattern completely constructive. At all other points in the image plane, light from any given point in the diffraction pattern arrives exactly half a wavelength out of phase with light from some other point in the pattern, and the net result is complete destructive interference.

Real objectives necessarily have a finite aperture and capture only a proportion of the light from the object. Light from the truncated Fraunhofer diffraction pattern still interferes constructively in the centre of the image plane, but because part of the diffraction pattern is 'missing', at lateral points in the image destructive interference is to a greater or lesser extent incomplete. The less light from the object is included in the Fraunhofer diffraction pattern, the broader is the image; if only a very narrow beam of light is admitted by the objective the image of a point object is spread out evenly over the whole image plane.

The image of a point object, formed by a perfect (aberration-free) lens of finite aperture, is the *Airy disc* first described in connection with astronomical telescopes (Airy, 1828). Viewed in two dimensions ('face-on') the Airy disc consists of a bright circular region surrounded by a number of paler secondary rings separated by dark rings. In cross section, a bright central peak is seen with the intensity dropping to zero a short distance away on both sides; further out still a series of peaks of decreasing intensity are present (Exercise E3). Figure 3.5 and Figure 3.6 show respectively amplitude and intensity plots of a one-dimensional scan computed with the Zernike program using an objective NA of 0.125 and distances along the horizontal axis marked every two wavelengths. The amplitudes of the central peak and the first lateral subpeaks are of opposite sign (remember that intensity is the square of the amplitude, and is always positive). Application of integral calculus (fairly elemen-

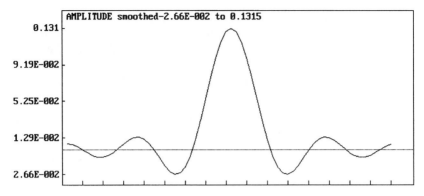

Figure 3.5   Airy disc (amplitude plot)

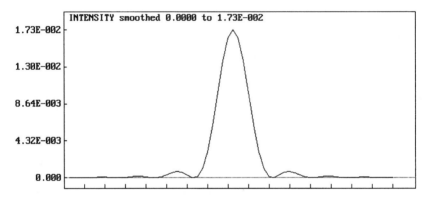

Figure 3.6   Airy disc (intensity plot)

tary but nevertheless beyond the capability of the author!) shows that in the relatively simple case of a rectangular objective aperture, as simulated by the computer program, the distance from the centre of the disc to the first dark ring is $0.5\lambda/NA$ (ignoring magnification due to the objective). If the objective aperture is circular, as is normally the case in practice, the corresponding distance works out as $0.61\lambda/NA$ (Longhurst, 1973, pp. 241 and 245).

The light emitted by a single point object is coherent irrespective of the nature of the illumination, but the imaging of two or more point objects is markedly affected by the coherence or otherwise of the light they emit.

3.8   INTERACTION OF COHERENT AND INCOHERENT LIGHT WAVES

The amplitude and phase of a wave can be represented respectively by the length and the direction of a vector, and the resultant of the interference of two coherent waves can be obtained by combining their vectors. Let A and B respectively be the amplitudes of two coherent waves which are $\theta/360°$ wavelengths out of phase with each other. The intensity of the resultant wave equals the square of its amplitude R, and is readily obtained by applying the cosine rule to the parallelogram of forces (Figure 15.1):

$$R^2 = A^2 + B^2 + 2.\cos(\theta).AB \qquad\qquad 3.6$$

Waves coming from different points of a self-luminous object, such as a fluorescent specimen illuminated with ultra-violet light, have however no constant phase relationship with each other, i.e. they are *incoherent*. We shall see (Chapter 5) that effectively incoherent illumination can also be obtained with conventional transmitted light using a condenser of high numerical aperture. Because the relative phase of different wave packets varies randomly and extremely rapidly from 0 to 360°, the cosine of the phase angle has a time-average of zero and the equation reduces to

$$R^2 = A^2 + B^2 \qquad\qquad (3.7)$$

The final intensity is then simply the sum of the contributing intensities, which is substantially different from that found with coherent interference. If the amplitudes (and hence also the intensities) of both A and B are unity, their interaction if incoherent gives light with an intensity of 2. If coherent, the resultant intensity may be anywhere between zero (if A and B are exactly out of phase so that $\theta = 180°$ and $\cos(\theta) = -1$) and 4 (if θ is zero and $\cos(\theta) = +1$). Energy is nevertheless conserved even in the coherent case, since constructive interference in one location is invariably accompanied by destructive interference elsewhere, and *vice versa*.

3.9   RESOLUTION OF TWO POINT OBJECTS WITH COHERENT AND INCOHERENT ILLUMINATION

According to Rayleigh (1896), two independent point objects emitting incoherent light of equal intensity may be considered to be resolved if the points are just far enough apart so that the centre of the Airy disc of one point falls on the first dark ring of the other. The combined image then has two symmetrical peaks separated by a trough, the intensity of which is about 26.5% lower than that of the peaks if the lens aperture is circular, and about 18.9% lower if the lens aperture is rectangular (Born and Wolf, 1975, p.415). According to this criterion the limits of resolution of two point objects emitting *incoherent* light are therefore about $0.61\lambda/$ NA and $0.5\lambda/NA$ respectively with a circular and with a rectangular aperture. If we accept that a drop of intensity of 26.5% in the centre of the combined image is necessary for resolution, the separation in the object plane of two point objects emitting *coherent* light is about $0.77\lambda/$NA (Born and Wolf, 1975, p. 424). Resolution is therefore significantly worse with coherent than with incoherent light. Note that using a rectangular aperture the limit of resolution on the Rayleigh criterion for two point objects ($0.5\lambda/$NA) happens to be the same as that for a grating object on the Sparrow criterion (see below), but in general the limit of resolution on the Rayleigh criterion for two point or line objects is *not* exactly the same as that derived by the Abbe method for a grating.

The Rayleigh criterion for resolution implicitly assumes a particular sensitivity of the eye or camera. With the advent of highly sensitive photodetectors such as CCD cameras and photomultiplier tubes, and with computer enhancement of images, a less arbitrary criterion due to Sparrow (1916) has become increasingly relevant. According to Sparrow's criterion, two objects may be considered to be resolved if the intensity half-way between the centres of their respective images is not greater than (i.e. is equal to or less than) that of either separate peak. For many types of object and illumination the Sparrow criterion gives a resolving power about 26% greater than does the Rayleigh criterion.

3.10 EFFECT OF OBLIQUE COHERENT ILLUMINATION ON THE RESOLUTION OF TWO POINT OBJECTS

If the illuminating light is a parallel beam of coherent light inclined at an angle to the optic axis, the light reaching a given object point will be to a greater or lesser extent out of phase with that reaching a neighbouring point. The Airy discs corresponding to the two object points will then have the same distributions of intensity as before, but because of their different phases their interaction will not be the same. The optimum phase difference from the point of view of resolution is half a wavelength, obtainable for two point objects one wavelength apart by illuminating light inclined at an angle corresponding to a condenser of numerical aperture 0.5. Under these conditions the two Airy discs are half a wavelength out of phase with each other, and at the midpoint between their peaks (where their intensities are equal but the amplitudes are of opposite sign) complete destructive interference occurs. The image therefore consists of two peaks separated by a zone of zero intensity, and the objects are 'resolved' with *any* finite objective aperture (Goldstein, 1992). The distance apart of the image peaks may however be different from that of the two objects, so that although the objects are nominally resolved a quite incorrect impression of the specimen may be obtained.

Historically, the effect of oblique illumination on the imaging of periodic or isolated objects has been commented on many times. Quekett (1852) noted that for '... the perfect definition of the markings of certain animalcules of the genus *Navicula*, very oblique light must be employed ...'. Töpler (1866), using a Schlieren method with a 'slider' in the back focal plane of the objective or eyepiece, observed but was unable to explain the fact that resolution of the fine lines of *Navicula angulata* was lost if the lines were parallel to the edge of the slider, while cross lines at right angles to this appeared surprisingly clearly.

The effect of oblique illumination on pale or transparent objects is somewhat similar to that of differential interference contrast (see Section 7.1).

3.11   A NOTE ON FRAUNHOFER AND FRESNEL DIFFRACTION PATTERNS

As already stated, the pattern of light visible in the back focal plane of the microscope object is a Fraunhofer diffraction pattern (see e.g. Jenkins and White, 1957, p. 288). This pattern is optically in a sense at an infinite distance from the object, since the rays which form it leave the object plane parallel to each other and are only brought together by the focussing action of the objective. In any event, the Fraunhofer pattern is a special case of diffraction, and diffraction patterns present in other situations (e.g. quite near the object) are in general called Fresnel patterns. A Fresnel diffraction pattern which is very near the object will of course closely resemble the light emerging from the object; as the distance from the object increases, the Fresnel pattern may undergo a number of complicated changes. When it is a very long way from the object the Fresnel pattern gradually becomes indistinguishable from a Fraunhofer pattern. The analytical mathematics of Fresnel diffraction is quite complicated, but the computer program can reproduce Fresnel phenomena reasonably well (subject to the usual reservations about one-dimensional Fourier transforms, failure to consider the obliquity factor, etc.).

To understand how Fresnel diffraction can be simulated, remember that an objective aperture of 1.0 in the program (physically unobtainable with a 'dry' lens) forms a perfect image of an in-focus object. If the objective is thrown out of focus by an arbitrary distance (say n wavelengths) a perfect image is still formed – but of a plane n wavelengths away from the object, containing a Fresnel diffraction pattern. The procedure to form a Fresnel diffraction pattern is as follows. Define an object in the usual way, carry out the Fourier transform, instruct the program to change the focus by an arbitrary amount (it does this by adding or subtracting a phase difference to each point in the Fraunhofer pattern which is a function of the square of the distance of the point from the optic axis), and without limiting the objective aperture carry out another Fourier Transform. The resulting image is a Fresnel diffraction pattern.

Several exercises demonstrating features of Fresnel and Fraunhofer diffraction patterns are given in Chapter 17.

3.12   OUT-OF-FOCUS IMAGES – DECONVOLUTION AND THE BECKE LINE

The diffraction-limited image of a point object extends both above and below the plane of best focus in a complicated pattern rather like a three-dimensional Airy disc – see e.g. diagrams in Born and Wolf (1975, p. 440) and Martin (1966, p. 91). Away from the plane of best focus, the image of a point changes from a bright disc surrounded by a series of alternating black and bright rings to a wider, less bright disc with a dark centre. At greater distances from best focus a bright spot again appears in the centre of the image and also expands into rings.

We have already discussed lateral resolution, i.e. the distance apart of two objects, the images of which can just be resolved in the plane of best focus. *Vertical resolution* can similarly be defined, explicitly or implicitly (see e.g. Wilson, 1990; Kimura and Munakata, 1990), as the smallest distance apart of two overlapping objects which can be shown by careful focussing to be separate. Both types of resolution depend on the objective NA, the mode of illumination, the type of object, and the criterion (e.g. Rayleigh or Sparrow – see Section 3.9) by which resolution is judged.

Wilson (1990), Kimura and Munakata (1990) and others discuss both lateral and vertical resolution in terms of the full-width-half-maximum ('FWHM') of the instrumental response, defined as the distance between the places in the image of a point object at which the intensity is half that at the maximum. This is equivalent to using the Sparrow criterion to determine the resolution of two incoherent point sources of equal intensity, since at any greater distance apart such objects will have a combined image the centre of which shows a finite dip in intensity.

An alternative approach to out-of-focus images is in terms of the *depth of field*, i.e. the greatest deviation from best focus which still gives an acceptable image of a defined object . 'Depth of field' of course depends on what percentage decrease in intensity is still regarded as acceptable; by analogy with the Rayleigh criterion this is often taken to be about 20%.

In the case of the imaging of a point object by a round aperture, mathematical analysis (Born and Wolf 1975, p. 441) shows that intensity along the optic axis decreases to (a first) zero at about $\pm 2\lambda/NA^2$, and to about 80% of the maximum at a distance of approximately $\pm 0.5\lambda/NA^2$. Using the present computer model, zero intensity is not reached in the centre of the image on defocussing, and the amount of defocussing necessary to give a 20% decrease in central intensity is not identical with that predicted by the equations of Born and Wolf. These discrepancies are probably due to the fact that the computer model is one- and not two-dimensional (for the assumptions and inherent limitations of the program see Section 15.7).

Depth of field on the object side of the objective should not be confused with *depth of focus* on the image side. In the case of a high-magnification objective with a large numerical aperture, the depth of field is very small but the depth of focus is large. This means that the position of the eyepiece is not very critical, and that the position of the film plane in a photomicrographic camera can be moved slightly without markedly affecting the sharpness of the picture. The opposite is found with a low-power objective with a small numerical aperture; focussing of the object is not very critical, but the depth of focus of the image is small. When adjusting the position of the eyepieces in a binocular body, therefore, one should first focus a suitable specimen using a high-power objective, and then (without changing the fine adjustment of the microscope) swing a low-power objective into place. The eyepieces should then be adjusted so that each gives a sharp image. This setting of the eyepieces will be suitable also for the high-power objective, since this has a greater depth of focus on the image side.

3.12.1 Deconvolution. With a thick object, the image at any given plane of focus is in general affected by out-of-focus detail of parts of the object in other planes. One approach to obtaining an improved image is confocal scanning microscopy (see Chapter 12). Another is to capture (say using a CCD camera and computer-controlled automatic focussing) a series of images at different focal planes. Using various *deconvolution* algorithms it is possible to use these images to compute an image more or less free from instrumental errors and out-of-focus detail (see e.g. Fay *et al.*, 1989; Macias Garza *et al.*, 1989). Several computer programs for this purpose are available commercially, and give considerable improvement of some images. Deconvolution is in principle compatible with confocal scanning microscopy, so that combining the methods can sometimes give results superior to that of either method used alone.

3.12.2 The Becke Line. If a microscope with transmitted axial illumination of low numerical aperture is focussed accurately on air bubbles or oil droplets suspended in water, and then focussed upwards slightly, a bright line at the edge of the object appears to move into the medium with the higher refractive index. Conversely, on focussing downwards the bright line moves into the medium with the lower index. It is thus possible to compare the refractive index of an object with that of the immersion medium (Johannsen, 1914; Saylor, 1935, 1966; Hartshorne and Stuart, 1970), although more modern methods of microscopy such as phase contrast are more sensitive for this purpose.

The rather obvious explanation for the focussing effect is that when suspended in water, air bubbles and oil droplets act respectively like negative and positive lenses (Dippel, 1872; Naegeli and Schwendener, 1877; Carpenter and Dallinger, 1891). Similar behaviour however occurs at vertical interfaces in thin sections, and this is not so readily explicable by ray optics. Salomon (1896) named the phenomenon the Becke line after Becke (1893), according to whom the phenomenon is due to total internal reflection of rays which are not accurately parallel to the optic axis. Hotchkiss' (1910) somewhat different ray-optical explanation also assumed slightly oblique light and a vertical interface, while Grabham (1910) discussed the possibility of light precisely parallel to the optic axis striking a slightly inclined interface.

Anything which can be explained by ray optics must in principle be consistent with diffraction theory. A role for diffraction in the formation of the Becke line appears to have been first suggested by Spangenberg (1921), on the basis of certain experimental observations. Kinder and Recknagel (1947) applied quantitative diffraction theory to out-of-focus images, equivalent to Fresnel diffraction patterns, of the edges of extended objects which were retarding and only slightly absorbing. With a semi-infinite object the general form (but not the size) of the diffraction pattern is essentially unaffected by its distance from the object. On focussing upwards the maximum intensity lay on the object side of the edge with object retardations less than half a wavelength, while with a retardation between half and one wavelength the maximum moved to the side away from the object. In the case of an opaque object or one with a retardation of precisely half a wavelength, the diffraction pattern was essentially symmetrical about the geometrical edge.

Faust (1951, 1955) extended the diffraction analysis to include the out-of-focus images of transparent strips of finite width. Relatively narrow strips are unlike semi-infinite objects in that the diffraction patterns from their two sides may interact with each other, so that the extent to which the image is out of focus is important. Faust's calculations suggested that with coherent illuminating light parallel to the optic axis, differences in the above- and below-focus images (i.e. the Becke effect) would be seen with objects with curved surfaces, but not with objects having straight edges parallel to the optic axis. However, Faust's choice of an object retardation of half a wavelength for his calculations was extremely unfortunate. It can be shown both experimentally and by using the Zernike computer simulation that flat retarding objects with vertical edges show Becke lines even with perfectly axial coherent illumination, *except* in the special case where the retardation is precisely half a wavelength. Plane objects with vertical sides show much more diffraction at their edges than do curved objects having the same central retardation because the change in phase at the edge of plane objects is more abrupt, and with small retardations (or advances) the Becke line effect is much better seen with plane objects. In the special case of a half-wavelength retardation or advance (which are of course equivalent), a plane object illuminated with coherent or almost coherent light shows marked diffraction fringes near its edges, but these change symmetrically on both sides of the plane of best focus so there is no Becke line effect.

3.13 APPLICATION OF THEORY TO PRACTICE – EFFECT OF THE SENSITIVITY OF THE HUMAN EYE

The effect on resolution of the sensitivity of the human eye has been investigated experimentally by Berek (1927, quoted by Martin, 1966, p. 201). His results suggest that the sensitivity of the eye to intensity is somewhat greater than assumed by the Rayleigh criterion, but that the eye can only distinguish objects subtending angles greater than about 4.7 minutes of arc, which is about twice as great as the usually cited figure of 2–3 minutes. Martin comments that these figures '... are the best comment on the very speculative character of the theory ...'.

Chapter 4

Extension of the ABBE Theory to Transparent Objects (Zernike's Phase Contrast)

The human eye is sensitive to differences in light intensity, but not to the phase of light. A transparent object with a refractive index different from that of the surrounding medium changes the phase but not the intensity of the light passing through it, and a perfect microscopic image of such an object is likely to be invisible. The visibility may however be improved to some extent by almost any manipulation of the image-forming process, for example by throwing the specimen slightly out of focus or by eliminating some diffracted light by the use of an objective of relatively limited numerical aperture. The *phase contrast* method of Zernike is an elegant and rather specific way to render transparent specimens visible. As already discussed (p. 28), in the Fraunhofer diffraction pattern of a coherently illuminated amplitude grating there is a phase difference of half a wavelength between the direct and the first-order diffracted beams, provided an (opaque) bar is on the optic axis. Zernike (1934) recognised that the corresponding phase difference with a pure phase object is only *one quarter* of a wavelength.

Consider the diffraction of light at a grating consisting of transparent bars separated by slits of the same width as the bars. If the refractive index of the bars is higher than that of the medium, light passing through the bars is relatively retarded by a phase angle p. A simple geometrical analysis (Figure 4.1) reveals that the direct light from the

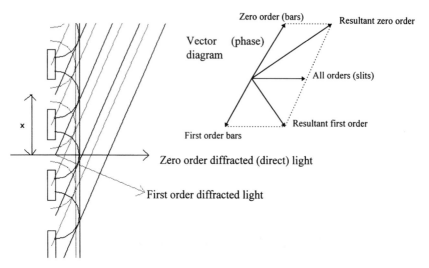

Figure 4.1 Diffraction at a phase grating

bars interacts with that from the slits to give a resultant somewhat retarded relative to the original direct light. The first-order diffracted beams from the bars and slits similarly interact, but their resultant is under the defined circumstances exactly one quarter of a wavelength (90°) out of phase with the resultant direct light.

Zernike's phase contrast method consists of introducing a *further* quarter-wave phase difference between the beams by inserting a suitable *phase plate* in the back focal plane of the objective, where the direct and diffracted beams are spatially separated. The diffraction pattern of the transparent object is now similar to that of an absorbing specimen, and the interference of the direct and the diffracted light results in a visible image.

4.1 PRACTICAL PHASE CONTRAST

In commercial phase-contrast microscopes the illumination is not in general an axial coherent beam, which would limit the resolution. Instead, a hollow cone of light comes from an annular opening in the front focal plane of the condenser, while the phase plate in the back focal plane of the objective has a slightly thicker or thinner ring which is conjugate with the image of the condenser annular aperture. The direct (undiffracted) light passing through this ring is advanced or retarded in phase, usually by about a quarter-wavelength relative to the diffracted light, most of which passes through the central or lateral zones of the phase plate.

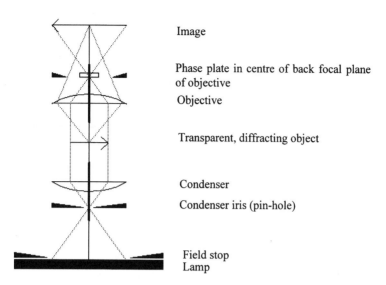

Image

Phase plate in centre of back focal plane of objective

Objective

Transparent, diffracting object

Condenser

Condenser iris (pin-hole)

Field stop
Lamp

Figure 4.2   Zernike's phase contrast with axial coherent illumination

In so-called 'dark' or 'positive' phase contrast the phase of the direct light is relatively *advanced* by a quarter wavelength, and a moderately retarding transparent object appears relatively dark on a bright background. Conversely, in 'bright' or 'negative' contrast the direct light is *retarded* by a quarter wavelength, and the image appears bright on a dark background (see the note below on terminology).

In many cases of practical biological interest the transparent object is thin, and its refractive index is not very different from that of the surrounding medium. Under these circumstances the intensity of the diffracted light tends to be much less than that of the direct light, and their mutual destructive interference to form the amplitude image is incomplete. The image contrast can be increased if the direct and diffracted light are made more equal in intensity, by absorbing some of the direct light in a layer of metal or carbon superimposed on the ring in the phase plate. This is the situation in the majority of commercially available phase contrast objectives, but the use of such a conventional ('A-type') phase plate is at the possible risk of a misleading representation of relatively retarding objects, which can anomalously appear brighter than thinner ones (*'reversal of contrast'*). If the specimen is unusually thick or highly retarding, the intensity of the diffracted light may actually be greater than that of the direct light, and to obtain the strongest contrast requires the absorption of some diffracted light by a ('B-type') phase plate in which the absorbing layer covers the whole phase plate *except* the ring through which the direct light passes.

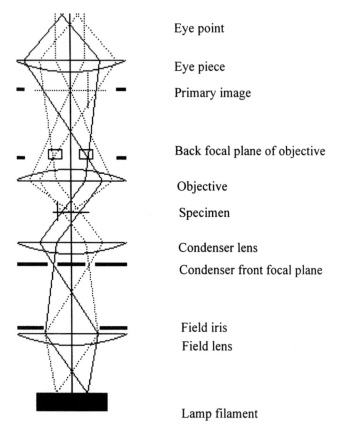

Eye point

Eye piece

Primary image

Back focal plane of objective

Objective

Specimen

Condenser lens

Condenser front focal plane

Field iris

Field lens

Lamp filament

Figure 4.3   Practical phase-contrast microscope

4.2   TERMINOLOGY OF PHASE PLATES

Bennett *et al.* (1946) defined A type phase plates as those '... in which the ratio of the intensity of the undeviated bundle to that of the deviated bundle is reduced ...'. Conversely, B plates absorb some of the diffracted light. If '.. the diffraction plate increases the optical path of the undeviated bundle relative to that of the deviated bundle, the letter A or B is followed by a + sign ...'. In other words, + plates retard the phase of the direct light. It was found empirically that with specimens having small differences of optical path in general '... the A(+) type of diffraction plate made details in the specimen brighter and the B(−) type darker than their surrounds and these will be called respectively bright and dark contrast ...'. 'Positive' phase contrast may be regarded as a synonym for 'Dark', but is *not* consistent with the + and − terminology of Bennett.

The fast Fourier transform used in the program has been slightly modified to give a correctly oriented image after only two transforms (see

p. 28). To do this the final imaginary components are reversed in sign, so that the phases displayed after a single transform are not quite as predicted by theory. The phase contrast routine however takes this into account so that the image given by a given type of phase plate is correct.

4.3 A SIMPLE THEORY OF IDEAL PHASE CONTRAST

If a uniform transparent object precisely fills half the microscopic field, as is the case with a pure phase grating with bars and slits of equal width (Figure 4.1), the amplitudes and phases of the direct and diffracted beams can be obtained by a simple vector method (Goldstein, 1982a), from which it is possible to calculate the apparent intensities of the object (bar) and background (slit) with a phase-plate of arbitrary retardation and absorption. Identical relationships, consistent with the conservation of energy, are yielded by an analytic method (Bennett *et al.*, 1946). Significantly different equations published by Barer (1952) and others explicitly or implicitly assume that the size and/or retardation of the object are very small and are inconsistent with the conservation of energy for large or highly retarding objects.

Figure 4.4 and Figure 4.5, based on Figs 7 and 8 of Goldstein (1982a), show the calculated apparent intensities of transparent, retarding objects which half fill the microscopic field, using phase plates which advance or retard the direct light by precisely one quarter wavelength and which have various transmissions (PT). It is assumed that the phase plate affects the direct but not the diffracted light, i.e. the phase contrast is 'ideal'. In Figure 4.4 the intensity of the object (grating bar) is plotted against the initial intensity of the illuminating light, while Figure 4.5 shows the intensity of the object (grating bar) vs. that of the background (grating slit). To read the apparent intensity of the background with a given object and phase plate, use the abscissa corresponding to the opposite sign of phase plate. For example, to read the intensity of the background with an A+ phase plate use the markings given for an object using an A− plate.

In ideal phase contrast using a quarter-wave retarding or advancing phase plate of any or no absorption, the apparent intensity of an object with a retardation of half a wavelength is identical with that of the background, so that the object will be invisible. With a non-absorbing quarter-wave A− or B− phase plate (PT = 1.0), the intensity of the image decreases to zero as the retardation of the object increases from zero to 90°, and then increases to equal that of the background so that the image is invisible, when the object retardation reaches 180°. At higher object retardations the object appears brighter than the background. The higher the absorption (the lower the transmittance PT) of the phase plate, the greater the sensitivity, that is to say objects of relatively small

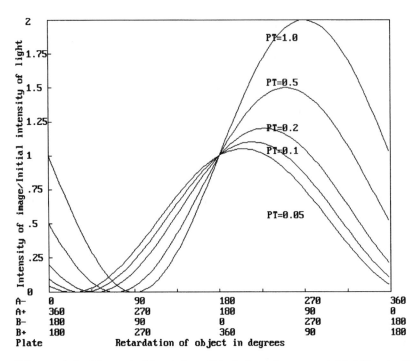

Figure 4.4   Theoretical intensity of ideal phase image vs. initial illumination

retardation appear black, but reversal of contrast may occur with even moderately retarding objects. Commercial phase plates are usually of the A– type with a transmittance of about 0.2 (i.e. 20%) for the direct light and give satisfactory dark contrast with many biological specimens. Some microscopists however prefer the 'negative' contrast given by an A+ plate, with which reversal of contrast is less of a problem. If you study Figure 4.5 carefully you will see that with a phase-plate of a given transmission, a negative-contrast (A+) plate gives reversal of contrast only with substantially more retarding specimens than is the case with a positive (A–) plate. As already stated, highly retarding specimens such as whole embryos may require a B+ or a B– plate. Very thin specimens such as sections prepared for electron microscopy are best studied with highly absorbing phase plates with a transmission of around 5% or less, but these are unfortunately not readily obtainable commercially.

4.4   A MORE GENERAL THEORY OF IDEAL PHASE CONTRAST

Let W be the object width as a fraction of the field width, OPD be the object retardation in radians, and T be the object transmittance. It can be

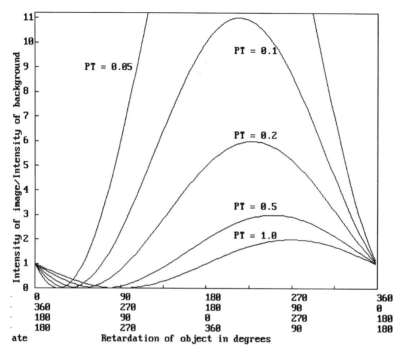

Figure 4.5   Theoretical intensity of ideal phase image vs. background

shown (Goldstein, 1991b) that the intensity I_0 of the zero order diffracted light is then

$$I_0 = W^2 T + (1 - W)^2 + 2W\sqrt{T}(1 - W)\cos(OPD) \qquad (4.1)$$

The phase angles A_0 and A_1 of the zero-order and diffracted light respectively are given by

$$\cos(A_0) = \frac{1 - W + W\sqrt{T}.\cos(OPD)}{\sqrt{W^2.T + (1 - W)^2 + 2W\sqrt{T}(1 - W)\cos(OPD)}} \qquad (4.2)$$

and

$$\cos(A_1) = \frac{\cos(OPD)\sqrt{T} - 1}{\sqrt{T + 1 - 2\sqrt{T}.\cos(OPD)}} \qquad (4.3)$$

In the case of a black object ($T = 0$) of width W the phase difference between the zero-order and diffracted light is 0.5 wavelengths, the intensity of the zero-order light $= (1 - W)^2$, and the total light intensity $= 1 - W(1 - T) = 1 - W$. The ratio of the direct to the total light therefore equals $1 - W$.

In order to obtain a black phase contrast image of a transparent object the relative phases and intensities of the direct and diffracted light on leaving the phase plate must resemble those of a black object, i.e. the phase difference between the zero-order and diffracted light must be 0.5 wavelengths, and the ratio of the intensities of the direct and diffracted beams be $(1 - W)$. In the general case the transmittance of the phase plate required to give a black image is

$$PT = \frac{(1 - W)^2(1 + T) - 2\sqrt{T}.\cos(OPD)}{W^2T + (1 - W)^2 + 2\sqrt{T}(1 - W)\cos(OPD)} \tag{4.4}$$

The required phase change in the zero-order light works out as

$$PR = A_1 - A_0 \pm 0.5 \tag{4.5}$$

Since it is impossible to *increase* the absolute intensity of the zero-order light, if the calculated value of PT is greater than unity it is necessary to decrease the intensity the diffracted light by multiplying by $1/PT$.

In the special case of a transparent object half filling the field ($T = 1$, $W = 0.5$), the above equations reduce to

$$PT = \tan(OPD/2)^2 \tag{4.6}$$

and

$$PR = -0.25 \tag{4.7}$$

These expressions were previously derived by Bennett *et al.* (1946) and Goldstein (1982a).

Table 4-1

Predictions of vector theories of phase-contrast	Object-width negligible	Object-width = half field
Retardation PR of phase-plate giving zero image intensity	$OPD/2 \pm 90°$	$\pm 90°$
Transmission PT of phase-plate giving zero image density	$4.[\sin(OPD/2)]^2$	$[\tan(OPD/2]^2$
Transmission PT of phase-plate (PR = $\lambda/4$) giving minimum image density	$[\sin(OPD)]^2$	$[\tan(OPD/2)]^2$
Intensity of this image relative to background	$[\tan(OPD/2)]^2$	0
Retardation of object giving reversal of contrast (PR = $\lambda/4$)	$OPD > 2.\arctan(PT^{1/2})$	$OPD > 4.\arctan(PT^{1/2})$
Retardation of object giving maximum reversed contrast (PR = $\lambda/4$)	$OPD = 180° + \arctan(PT^{1/2})$	$OPD = 180° + 2.\arctan(PT^{1/2})$

If W tends to zero, i.e. the object is negligibly small relative to the microscopic field,

$$PT = 4.\sin(OPD/2)^2 \tag{4.8}$$

and

$$PR = -0.25 + OPD/2 \tag{4.9}$$

as stated by Barer and others. Previously published equations are therefore special cases of the general equations given above.

4.5 NON-IDEAL PHASE CONTRAST

The above equations relate to ideal phase contrast, in which the phase plate affects all the direct light and none of the diffracted rays. In practice at least the lowest orders of diffracted light are likely to be influenced by a central phase plate of finite diameter, or by a phase annulus of finite width. In general this results in the phase contrast effect being limited to the edges of extended objects, and the appearance of an adjacent phase 'halo' of opposite contrast. Thus in 'dark' phase contrast the centre of an extended specimen appears similar to the background, its edges appear darker than the background, and a bright 'halo' is seen in the immediately adjacent background. Converse effects are seen in 'bright' phase contrast. These effects are simulated in the program by specifying a finite width of phase plate.

It is also possible with the program to simulate the effect of partially coherent illumination on phase contrast, using either axial illumination or oblique illumination coming symmetrically from both sides of the optic axis (see p. 134).

4.6 VISUALISING PHASE OBJECTS WITH ORDINARY BRIGHT-FIELD MICROSCOPY

In 1942 Zernike compared phase contrast qualitatively with the Schlieren method, oblique dark ground and central dark ground illumination. He also explained satisfactorily for the first time the well known fact that the visibility of a pale or transparent object is improved in conventional transmitted-light microscopy by almost closing the iris diaphragm of the condenser and thus illuminating the specimen with a narrow central pencil of light. Zernike pointed out that when doing this it is usually necessary also to throw the specimen a little out of focus, which is equivalent to moving the image plane slightly and introducing a difference in the optical paths to the image from the direct light (in the centre of the back focal plane of the objective) and from the diffracted light (which mainly passes through more lateral parts of the objective back focal plane). The essential condition for Zernike's phase contrast has been met if this introduced phase difference is about a quarter of a wavelength. If the transparent object is thicker than the depth of focus of the objective it may not be necessary to defocus the object, since at least part of the object will probably lie at the required level and thus be rendered visible.

The 'phase-contrast' image obtained by the mechanism just described is naturally deficient in several respects. The small condenser aperture makes the resolution poor, the relative intensity of the direct light is not reduced as is the case with an orthodox phase plate so that the contrast is not optimal, and the image is of course out of focus. An imperfect, blurred but visible image is nevertheless preferable to one which is theoretically perfect but invisible, and much important work in cell biology was carried out by this simple means before the invention of phase contrast.

Reducing the condenser aperture may by an analogous mechanism improve the visibility of pale absorbing specimens, at the cost of resolution. Often a condenser aperture of about two-thirds or three-quarters of the objective aperture proves most satisfactory in practice, and this is the setting of the condenser iris conventionally recommended to beginners in microscopy. There is however no universally ideal setting for the condenser aperture, and the best compromise between resolution and contrast depends on the nature of the specimen and the type of information sought.

4.7 SELECTED HISTORICAL NOTES

Although phase contrast was invented by the Dutch physicist Frits Zernike around 1932 (Zernike 1934, 1942), his work was not immediately appreciated. When he '... approached Zeiss in 1932, one of their staff said "If this had any practical value, we would ourselves have invented it long ago ...".' (Barer, personal communication 1966).

Apparently no worker before Zernike noticed that the diffraction theory necessary to understand phase contrast had long been available in the work of Abbe (Zernike, 1934), but Zernike was neither the first nor the last to consider the imaging of phase objects. The extensive literature cannot be reviewed here (for some references see Menzel, 1957; Wolter, 1950a, b, c and d and Goldstein, 1982a and 1991b), but perhaps a tribute may be paid to some important early work which has largely been forgotten.

Foucault (1858) and Töpler (1866), both of whom were cited by Zernike, studied the subject in some detail but interpreted their work only in terms of geometrical optics. Stephenson (cited by Hogg, 1883) recognised that the visibility of transparent objects depends on a difference between the refractive indices of object and medium. Hogg himself believed the visibility of a transparent object to be directly proportional to this difference, and therefore thought that a diatom with a refractive index of 1.43 would be six times more visible mounted in phosphorus (RI 2.11) than in balsam (RI 1.54). We now know from the work of Zernike that the relationship of contrast to refractive index is not quite as simple as this.

Exner (1887) explicitly stated the importance of diffraction in the imaging of unstained striated muscle fibres: 'When a plane wave traverses discs of unequal refractive index, it acquires parallel ridges corresponding to the layers of smaller index. The problem then consists in the determination of the resultant of the interference of the elementary waves proceeding from a surface of this form ...'. In discussing work by Merkel (1873) Exner wrote that if '... the fibrilla is only visible in blue light upon a dark field the difference in path of the two rays must amount to $\lambda/2$ ...'. Exner apparently did not appreciate that a much smaller path difference might still be visible, even if the image is not completely black, and he curiously suggests that if a microscopic appearance is altered with oblique illumination '... it indicates that a complete representation is not obtained by central illumination, and it must be doubtful whether it is so by oblique illumination ...'. If this criticism were valid, any improved method of microscopy would immediately and automatically be invalidated!

In highly original work which clearly anticipated phase contrast, apodization and 'Gegenfeld', but which was almost completely ignored by later workers including Zernike, Bratuscheck (1892) described (among other phenomena) the effects on a microscopic image of attenuating either the direct or the diffracted light with a thin layer of platinum or carbon in the back focal plane of the objective. He noted that a carbon film changed the phase as well as the intensity of the affected light, and that absorbing some of the direct light gave stronger images of object details which had only small differences in transmittance or retardation.

Chapter 5

Function of the Microscope Condenser – Partially Coherent Illumination

A simple (*critical*) microscope illumination system typically consists of a light source such as an incandescent filament, an iris diaphragm (field-stop) at or very near the light source, and a condenser lens with an iris diaphragm (aperture-stop) in its front focal plane (Figure 1.3). An image of the light source is formed by the condenser in the plane of the specimen; irregularities in the source such as a visibly coiled filament may be smoothed out in its image by a diffuser (e.g. a piece of ground glass) situated between the lamp filament and the field-stop. In the *Köhler* system of illumination (Figure 1.2) the uniformity of illumination in the specimen plane is improved by an additional lamp lens which acts as a uniform secondary source. An image of the lamp filament is then formed not in the object plane but in the front focal plane of the microscope condenser and also in the conjugate back focal plane of the objective. There is little or nothing to choose theoretically between critical and Köhler illumination as far as resolution and contrast are concerned, and both are suitable for all forms of microscopy (see also Chapter 1).

5.1 CORRECTLY ADJUSTED ILLUMINATION

The following facts about correctly adjusted microscopic illumination are fully explained in standard texts and are stated here without proof.

(1) The intensity of illumination of the object and image is directly proportional to the *intrinsic brilliance* (light output per unit area) of the source but is unaffected by the size of the source, which influences only the maximum area of specimen which can be illuminated. For example, if a 50W high-pressure mercury arc lamp is replaced by a 100W lamp with a larger arc but a lower intrinsic brilliance, the intensity of illumination is actually decreased.

(2) Stray light (glare) in the system, mainly due to multiple reflections at glass-air interfaces inside the objective, degrades contrast by spreading a haze of light over the image. It can be reduced by restricting the illumination to the part of the object being studied at the moment – this is the function of the field diaphragm.

(3) Reducing the numerical aperture of the condenser sometimes reduces glare slightly, probably because reflections tend to be greater if light strikes the highly curved surfaces near the edges of objective lens components.

(4) Provided the condenser entrance pupil (front focal plane) is filled with light from the source, the intensity of illumination of the specimen and the intensity of the final image are independent of both the distance of the source from the microscope condenser, and the magnification of the condenser.

(5) The intensity of illumination is proportional to the square of the effective condenser numerical aperture (but in the one-dimensional model used in the computer program it is *directly* proportional to the aperture). Closing the condenser diaphragm therefore reduces the intensity of the illumination, but this is an incidental effect – the intensity of illumination should be controlled by altering the power supply to the lamp or better still by inserting neutral absorbing filters in the optical path.

We have already seen (p. 32) that microscopic resolving power can be improved by using incoherent instead of coherent illumination, and that the visibility (contrast) of transparent or pale specimens can also be improved by almost closing the condenser iris and simultaneously throwing the specimen slightly out of focus (Section 4.6). We shall now discuss the practical use of the condenser iris (aperture diaphragm) in more detail.

5.2 FUNCTION OF THE CONDENSER APERTURE DIAPHRAGM (IRIS)

In the discussion on p. 31 it was assumed that coherent light is produced by a distant point source, and that incoherent light is emitted by a fluorescent object. In ordinary transmitted-light microscopy the coherence of the illumination and hence both contrast and resolution are in fact controlled by altering the diameter of the condenser diaphragm, which changes the angle of light illuminating the specimen. This is the primary function of the condenser.

In both critical and Köhler systems (Figure 1.2 and Figure 1.3) effectively coherent illumination is obtainable by almost closing the condenser diaphragm. Under these circumstances the pin-hole aperture acts as a source of coherent light for the whole object, which is illuminated with effectively parallel beams of light (Figure 5.1).

Another way of thinking about this is to consider how the condenser lens forms an image in the object plane of a given point in the (extended) light source. In Figure 5.1 the light source and the object are in conjugate planes, and one might suppose on the basis of geometrical optics that separate light-emitting points in the source would be imaged separately in the object plane. If however the condenser aperture is very small, the Airy disc of any given point in the source is spread out evenly over the whole object, which is therefore uniformly illuminated with coherent light. The same is true of every other point in the light source, so that the hypothetically separate images due to the separate, mutually incoherent points in the light source are identical, and the total illumination is effectively coherent.

Incoherent illumination could theoretically be obtained if the numerical aperture of the (dry) condenser tends to unity, corresponding to an (impossible!) illumination half-angle of 90°. Each light-emitting point in

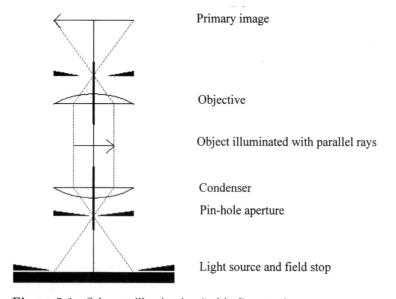

Primary image

Objective

Object illuminated with parallel rays

Condenser

Pin-hole aperture

Light source and field stop

Figure 5.1   Coherent illumination ('critical' system)

the source would then be imaged perfectly into the object plane, i.e. its Airy disc would be smaller than the limit of resolution of the microscope objective, and each resolvable part of the object would effectively have its own independent light source.

Complete coherence and complete incoherence are extreme, ideal cases. In practice the condenser aperture is necessarily finite but less than unity, and although the Airy disc corresponding to a given point in the light source is centred over a particular point in the object, some of the energy falls elsewhere. There is then some correlation between the phases of the light reaching and leaving neighbouring points, i.e. the illumination is *partially coherent*. The interaction of partially coherent light rays results in an image intensity somewhere between the values for perfectly coherent and perfectly incoherent light. No simple equation expresses such complex effects directly (Born and Wolf 1975, pp 522 *et seq.*), and in order to calculate the results each point in the light source must in principle be considered in turn. The intensity of the corresponding Airy disc in the object plane is computed taking into account the numerical aperture, glare, focus and primary spherical aberration of the condenser. The computed Airy disc falling on and around a particular object point is then modified by the absorption and retardation of the specimen, and its contribution to the final image is computed by performing a Fourier transform, modifying the Fraunhofer diffraction pattern in accordance with the defined properties of the objective, and carrying out a *second* Fourier transform. The whole process must be repeated N^2 times where N is the number of points in both the light source and in the object, and is accordingly relatively costly in terms of both time and computer memory (although one or two short-cuts may be made). The final image is obtained by simply adding the intensities of all the images attributable to individual, mutually incoherent points in the source.

Control of the coherence of illumination allows a trade-off between contrast and resolution, and it is probably true (as commonly taught) that with many objects the best condenser aperture is somewhat less than the aperture of the objective being used. The optimum condenser aperture however depends not only on the dimensions, transmittance and retardation of the specimen but also the sort of information sought. It can be shown theoretically that with two near-by absorbing line objects very similar (but not quite identical) results are obtained with completely incoherent illumination, and with a condenser aperture equal to that of the objective. Curiously, slightly *better* resolution of such objects is sometimes obtainable if the condenser aperture is somewhat greater than the objective aperture but less than unity (Hopkins and Barham, 1950; Goldstein, 1992) – this phenomenon is readily demonstrable with the computer program.

5.3   FOCUS AND CORRECTIONS OF THE CONDENSER

It might be thought that the coherence of the illumination would with a given condenser aperture be affected by the focus and spherical aberration of the condenser, since these would be expected to change the size and shape of the Airy disc corresponding to each light-emitting point in the source. This proves not to be the case (Martin 1966, p. 155). Provided the area of the light source is large enough so that any ray which emerges from the exit pupil of the microscope (the eye-point above the eye-piece) and which can be traced back through a point on the object can be traced further back to a source point of suitable luminance, the corrections and focus of the condenser do not influence the effective numerical aperture. Using a reasonably well-corrected and correctly adjusted condenser is nevertheless justified in practice, since only then can a small field-stop be imaged sharply into the plane of the object. This is necessary in order to limit as far as possible the area of specimen which is illuminated, and thus reduce to a minimum the amount of stray light which may eventually reach the eye or camera and degrade the image by 'glare'. In the computer program the condenser aperture, focus and spherical aberration have no effect on the imaging of the object unless the width of the field stop is reduced.

The *maximum* angle of light from the condenser which strikes the object is not the only factor which can affect image formation. Somewhat different results are obtained with a solid cone of illuminating light, which is normally the case with bright-field microscopy, and with a hollow cone of light such as is often used in phase contrast microscopy. The computer program can simulate both these types of partially coherent illumination, as well as partially coherent illumination coming from only one side (unilateral oblique illumination).

Chapter 6

Interference Contrast

The following account primarily concerns transmitted-light interference microscopy, but some of the discussion is applicable also to the incident-light variety which is used mainly for the examination of surfaces.

In both phase contrast and interference contrast microscopy a visible image of a transparent object is produced by the interference of coherent beams of light. In both cases these beams are spatially separated at some place in the optical system, where their relative retardation can be manipulated. In phase contrast the interfering beams are the direct (zero-order diffracted) light and the light diffracted by the object; which pass through different parts of the back focal plane of the objective. Interference contrast microscopy, on the other hand, does not inherently depend on diffraction although diffraction in practice influences the appearance of the final image; the two interfering beams are produced by some form of beam divider and are spatially separated in the plane of the object.

An idealised interference system might consist (Figure 6.1) of a source of coherent light A, beam-splitters B and D, totally reflecting mirrors C and F, and screens E and G.

The identical beam-splitters are dielectric (i.e. do not contain any metal, and absorb none of the light striking them), and each allows through half the intensity of the light striking it and reflects the other half. Some light reaches screen E along the path ABCDE and some along the path ABFDE. Since the beam-splitters each divide the light striking them into two beams of equal intensity, one would expect a quarter of the original intensity to reach screen E along each of these paths. If the light

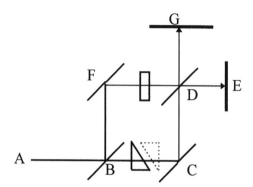

Figure 6.1   Idealised 'round-the-square' interference system

were completely incoherent and interference could therefore be ignored, half the original intensity would end up at screen E and similarly half at screen G. But since the light is coherent, interference cannot be ignored. Assume the original intensity from A is unity. Then the intensity at screen E due to light reaching it along path ABCDE is 0.25, and its amplitude (the square root of 0.25) is 0.5. So is the amplitude of the light reaching screen E along the path ABFDE. Before reaching screen E the light passing along the path ABCDE has undergone one total reflection (at C), has passed through one beam-splitter (B), and been reflected off one beam-splitter (D). Light reaching screen E along path ABFDE has similarly undergone one total reflection (at F), one reflection off a beam-splitter (at B) and one transmission through a beam-splitter (at D) Whatever changes in phase may have taken place at the mirrors or beam-splitters must have been identical for the two beams, and may therefore be ignored (remember that only relative changes in phase are significant). At screen E the two beams therefore are in phase with each other and interfere constructively to give a resultant of amplitude 1.0 (the amplitudes are additive). The corresponding intensity is 1.0 (the square of 1.0), and since the original intensity was also 1.0 and light energy is conserved, all the light from A must end up at screen E.

How can this be? Why is no light energy found at screen G? The answer to this is not immediately obvious, and is not discussed in most textbooks of optics (presumably because it is thought too obvious to the mathematical and too difficult for the rest of us!). We need to consider in slightly more detail the phase changes undergone by light reaching screen G. Light travelling along the path ABFDG undergoes one total reflection (at F) and two partial reflections (one at B and the other at D). Light travelling along the path ABCDG undergoes one total reflection (at C) but two partial transmissions (at B and D). The phase changes on total reflection (which happen to be exactly a half-wavelength each time) cancel out between the two beams and can be ignored, but the phase changes at the beam-splitters are not the same for the two beams since one beam experiences two partial reflections and the other two partial transmissions. Applying some rather complicated equations (Born and Wolf 1975, pp. 59–63) it turns out that each beam-splitter introduces a phase difference of precisely one quarter of a wavelength between the reflected and the transmitted light. There is therefore a total phase difference of half a wavelength between beams ABCDG and ABFDG, and on reaching screen G they undergo complete destructive interference so that no energy actually arrives at G.

We may note in passing that the relative phase change between light reflected from and that transmitted through a metallic beam-splitter such as a half-silvered mirror is almost but not quite a quarter wavelength (Born and Wolf, 1975, pp. 628–631). A small amount of energy is absorbed by the metal, and the law of the conservation of energy no longer applies strictly to the beams of light.

What happens if we now introduce a flat-sided retarding (but not absorbing) object between F and D (Figure 6.1)? In general the light reaching F along the path ABFDE is no longer exactly in phase with that travelling along path ABCDE, their interference is not perfectly constructive, the intensity of light at screen F is more or less reduced, and the 'missing' light energy turns up at screen G. The extent to which the light has been retarded by the object might in principle be estimated (within certain limits) by measuring the intensity at screen F and applying some mathematics, but it is more directly obtained by introducing a compensator into the system between B and C. This compensator might consist of a transparent wedge (solid lines in the diagram) plus a fixed wedge facing the other way (dotted in the diagram). The first wedge is pushed across the beam by rotating a screw, and when the intensity at screen F is restored to its previous value the retardation of the object equals that due to the movable wedge and can be read off the calibration of the screw. In this system the retardation is altered uniformly across the screen, but if there is only a single wedge a series of alternating dark and light fringes will be seen at F. A sort of contour map across the image of the object is seen, with bright fringes at places where the retardation of the object is compensated by exactly one wavelength, and dark fringes where the two interfering beams are half a wavelength out of phase.

Many interference systems have been used in microscopy, but those of greatest practical importance can be classified broadly into shearing and differential contrast types. The following brief descriptions ignore the imaging role of the microscope condenser and objective, both of which can substantially affect the resolution and quality of the image.

6.1   PRACTICAL INTERFERENCE CONTRAST: THE MACH–ZEHNDER SYSTEM

The Mach–Zehnder interference system is relatively easily understood, and is in essence the same as the idealised system described above with the simple addition of objectives and condensers into both beams. A beam of light is divided into two coherent beams by a beam-splitter (e.g. a half-silvered mirror or prism) which allows one beam to pass through, and reflects the other through a right-angle. The deflected beam is again deflected through a right-angle, this time by a totally reflecting mirror or prism, and both beams pass parallel through the object plane. One beam (the 'object' beam) passes through the specimen, and other (the 'reference' beam) through an empty area of medium. The previously undeflected beam is now deflected by a mirror, and both beams enter a second beam-splitter which acts as a beam uniter. Interference microscopes using this system have separate condensers and objectives for the specimen and reference beams, and have the desirable property of wide separation of the two beams in the plane of the specimen so that large specimens can be examined. They are however difficult and therefore expensive to manufacture with the necessary mechanical stability, and although once produced by Leitz none is currently commercially available.

6.2   PRACTICAL INTERFERENCE CONTRAST: THE JAMIN–LEBEDEFF SYSTEM

Interference microscopes of the Jamin–Lebedeff type exploit the properties of polarised light (see Chapter 14 and Figure 6.2) to split and combine the beams. Linearly polarised light from a polarizer (e.g. the proprietary dichroic material 'Polaroid') is divided in a birefringent calcite plate cemented to the top of the condenser into two beams which are linearly polarised at azimuths at right angles to each other. These diverge in the plate and are separated in the object plane by about 100–200 μm in the case of a low-power objective, and 30–50 μm if a high-power objective is used. The beams passing through the object and the empty background (respectively the 'object' and 'reference' beams) are reunited in a calcite plate attached to the front of the objective, pass through some type of polarising compensator and finally an analyser to reach the image. Objectives and condensers are manufactured in matched pairs with identical calcite plates cemented to their facing surfaces, and a half-wave retarding plate cemented to the top of the condenser reverses the azimuths of polarisation of the object and reference beams and enables the calcite plate attached to the objective to act as a beam uniter.

The relative phase of the object and reference beams and hence the contrast of the image can be controlled in various ways in a Jamin–

Lebedeff microscope. If a quartz wedge (see p. 106) is inserted below the analyser in a plane conjugate with both object and image the relative retardation of the object and reference beams is compensated by an amount which varies linearly along the wedge. This results in a series of fringes crossing the object plane, the separation of adjacent fringes corresponding to a retardation of one wavelength. The fringes are straight and parallel where they lie over empty background, but are displaced where they cross the edges of a phase-changing object by a distance proportional to its retardation. This produces a sort of contour map of the specimen from which its retardation can be estimated visually with a precision of about one tenth of a wavelength, over a range of several wavelengths. If monochromatic light is used relatively sharp fringes are seen over the length of the wedge, but with white light individual fringes can be distinguished – this is useful if the fringe displacement is more than one wavelength.

The Sénarmont compensator typically used in the Jamin–Lebedeff interference system (Figure 6.2) consists of a quarter-wave plate fixed with its slow axis at 45 to the transmission axis of the polarizer, and a rotatable analyser. It can be shown that the angle through which the analyser must be rotated to compensate for the retardation of an object is directly proportional to the retardation. Under favourable conditions, and with the aid of a half-shade eyepiece to facilitate precise setting of the

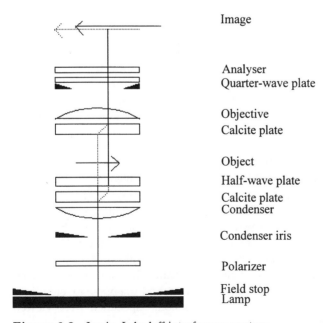

Figure 6.2 Jamin–Lebedeff interference system

analyser, the retardation of the object can be measured in a range of ± 1 wavelength with a precision of about one hundredth of a wavelength. Retardations of more than one wavelength cannot readily be distinguished with this method, and require preliminary examination with a quartz wedge compensator.

In all shearing systems the beamsplitter between the object and image planes results in the appearance of two images, one of which is called the 'real' image and the other the 'ghost' image. In order to avoid confusion the latter is sometimes deliberately made astigmatic and out-of-focus. Ideally these images are completely separate, but if the object is wider than the separation of the object and reference beams the real and ghost images partly overlap in the image plane, and measurements in the region of overlap become difficult or impossible.

6.3   PRACTICAL INTERFERENCE CONTRAST: DIFFERENTIAL INTERFERENCE CONTRAST

In differential interference contrast ('DIC') the separation of the object and reference beams in the object plane is very small, typically less than the resolving power of the objective. The 'real' and 'ghost' images therefore overlap almost completely, and the object beam interferes not with a reference beam passing through empty background as in the case of a shearing system, but with a reference beam which passes through an immediately adjacent part of the specimen itself. The result is that the contrast of the image is a function of the gradient of phase change in the specimen.

Figure 6.3 shows a DIC system in which a Wollaston prism in the front focal plane of the condenser splits plane polarised light from a polarizer into two orthogonally polarised beams. These leave the prism divergent, and are rendered parallel by the condenser. The beams pass through the object, are made convergent by the objective and are united by a second Wollaston prism in the back focal plane of the objective, and finally pass through an analyser.

Such a system is impractical with high-power objectives, in which the back focal plane is inaccessibly situated between elements of the lens. The well-known Nomarski system uses ingeniously modified prisms which can be located physically outside the lens; transverse movement of the upper prism alters the relative retardation of the object and reference beams, and permits the image contrast to be adjusted to give a bright image on a dark background, a dark image on a bright background, and intermediate appearances. Typically the contrast is adjusted so that one side of the specimen appears dark and the opposite side light ('shadow-cast'), almost as if a three-dimensional specimen were viewed with oblique illumination.

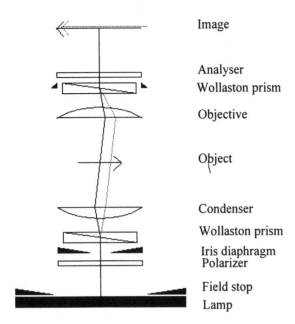

	Image
	Analyser
	Wollaston prism
	Objective
	Object
	Condenser
	Wollaston prism
	Iris diaphragm
	Polarizer
	Field stop
	Lamp

Figure 6.3   Differential interference contrast

Under most conditions the best results are obtained if the object and (displaced) reference beams are of equal intensity, but slight rotation of the polarizer permits the relative intensities to be adjusted. The smaller the numerical aperture of the condenser, the higher the contrast but the lower the resolving power – an optimum adjustment must be found by trial and error for each object. When used with a fairly large condenser aperture, DIC has the advantage over phase contrast of having a rather small depth of focus, which enables (say) the surface of a cell to be examined almost without disturbance from out-of-focus detail.

6.4   APPLICATIONS OF SHEARING INTERFERENCE MICROSCOPY

These cannot be gone into in detail here, but briefly it is possible with interference microscopy to measure the optical path difference (OPD) between an object and the mounting medium. The OPD equals the thickness T of the object multiplied by the difference between the refractive index u_0 of the object and the refractive index u_m of the medium:

$$OPD = Tu_0 - Tu_m = T(u_0 - u_m) \qquad (6.1)$$

In general the refractive index of the medium is known or can easily be measured, and if either the geometrical thickness of the object or its refractive index are known the other can be calculated. By making measurements successively in two or more media of different (but known) refractive indices, both the actual (geometrical) thickness of a specimen and its refractive index can be obtained. It is in theory possible (Goldstein, 1968) to obtain both the thickness and the refractive index of a flat object, mounted in a medium of known refractive index, by measuring the OPD using light inclined to the optic axis at different angles, but the precision and accuracy of microinterferometry is insufficient for the method to be useful in practice (Goldstein, 1994).

The refractive index of a solution is directly proportional to both the concentration of solute and to a constant called the specific refraction increment. The specific refraction increment is defined as the increase in refractive index if one gram of solute is added to 100 ml of solute, and its numerical value is approximately constant (0.0017) for aqueous solutions of most constituents of cells. Interference microscopy of living cells in water can therefore measure the density of the cells, expressed as mass per unit projected area. By making repeated measurements over the surface of the cell, which is readily done with an automatic scanning microinterferometer, it is possible to obtain the total dry mass of the cells, i.e. the total mass of the contained solutes.

6.5   COMPARISON OF DIFFERENTIAL INTERFERENCE CONTRAST WITH OBLIQUE ILLUMINATION

With oblique coherent illumination the phase of the light striking an object point is a function of its position in the object plane. With a finite objective aperture the Airy discs of adjacent image points overlap, and because of their difference in phase the interaction can result in an image of a transparent object closely resembling that of differential interference contrast, i.e. a 'shadow-cast' image dark on one side and light on the other. Although cheap and easy to obtain, such images tend to be non-uniform over the field of view, and because of the use of coherent (albeit oblique) illumination the resolving power varies substantially with the orientation of the object. Oblique illumination with video enhancement, for the visualization of transparent objects, is discussed by a number of authors including Bretschneider and Teunis (1994).

6.6   COMPARISON OF INTERFERENCE CONTRAST AND PHASE CONTRAST

Phase contrast can be regarded as a special case of interference contrast, but in practice there are pros and cons of both types of microscopy. Phase contrast is easy, quick and relatively cheap, and gives excellent contrast of thin, slightly retarding specimens. It is however essentially non-quantitative, and the halo, which is unavoidable in practice, can hamper observations of closely packed objects.

With shearing interference microscopy precise measurements can be made of suitable objects, and the images are free from the phase-contrast halo. Ghost images can however be troublesome, particularly with wide or closely packed objects, and the contrast with very thin specimens is usually not as good as with phase contrast. The apparatus is relatively expensive and difficult to adjust.

The images obtained with differential interference contrast are often very striking, and the method is popular both in biology and materials science. It is however difficult or impossible to quantify the results obtained, and three-dimensional appearances can be misleading. Thus a cell nucleus with a refractive index higher than that of the cytoplasm may incorrectly appear with DIC to project from the cell surface, and great care must be taken in the interpretation of images.

6.7 INCIDENT-LIGHT DIFFERENTIAL INTERFERENCE CONTRAST

Nomarski and similar DIC systems have been adapted for use with incident light, in order to study the surface of opaque specimens. The optical system is in principle similar to that illustrated for transmitted DIC, as used in biology, except that the optics are 'folded' so that a single lens serves as both objective and condenser. Such systems are extremely sensitive to small differences in height, and have been successfully used in metallurgy and in the electronics industry.

6.8 COMPUTER SIMULATION

The program permits the distance between the object and reference beams to be adjusted. A minimum half-wavelength separation gives DIC, and a shear greater than the width of the defined object gives shearing interference. The proportion of the total energy in the reference beam can be altered, corresponding to rotation of the polarizer in DIC, but should normally be set to give equal intensities of the reference and object beams. The default value of the phase difference between the object and reference beams is 0.25 wavelengths, which gives a conventional DIC image; altering the phase difference simulates transverse movement of a prism in DIC or rotation of the analyser in a Jamin–Lebedeff system. The program first displays the effect of the defined object on the transmitted light, then the 'ideal' interference image which would be obtained with an objective of unlimited aperture. The modification of the image by a finite objective NA (and possibly also spherical

aberration, focus error and stray light) is then simulated just as in the
case of ordinary bright-field illumination by application of the FFT,
manipulating the data appropriately, and applying the FFT for a second
time.

Chapter 7

Dark-field and Related Techniques

In dark-field (dark-ground) microscopy (Siedentopf, 1907b, 1909; Gage, 1920, 1925) the direct (zero-order diffracted) light is somehow blocked or intercepted, and the image is formed only by light refracted, diffracted or scattered by the object. Edges and discontinuities of both absorbing and phase objects appear bright on a dark background, and it is often difficult to distinguish between transparent objects of low and of high retardation, or between retarding and absorbing objects. The results are affected in rather subtle ways by precisely how the direct light is eliminated.

As already explained, in ordinary bright-field microscopy the image is formed by interaction of the direct light and at least the first order of diffracted light. In dark-field illumination the direct light plays no part in image formation, and for an image to be formed it is necessary for at least two different (and preferably adjacent) orders of diffracted light to be captured by the objective. This affects the maximum resolving power in complicated ways which depend not only on the particular dark-field system employed but also the precise nature of the object itself. Sometimes 'pseudo-resolution' is found, i.e. features appear in the image which do not correspond with structures in the object.

In some varieties of dark-field microscopy the numerical aperture of the illumination is greater than that of the objective. Either an objective of relatively low aperture must be used, or the objective aperture is reduced, for example by a built-in iris diaphragm or the insertion of a 'funnel' stop into the objective mount. This necessarily affects the possible resolution. In other types of dark-field microscopy the illumination is axial and more or less coherent, and the numerical aperture of the objective is unrestricted. Nevertheless, due to the elimination of the direct light the resolution is likely to be poorer than with coherent bright-field microscopy using an equivalent objective, and substantially worse than that of bright-field microscopy with partially coherent illumination (a finite condenser aperture). Due to the marked increase in contrast, reduction in resolution may not however be obvious with dark-field microscopy.

The oldest and probably simplest system is *oblique illumination*, in which a parallel beam of light strikes the specimen at so steep an angle relative to the optic axis that none can enter the objective directly. This method is not often used nowadays, but has some interesting properties (see below). In ordinary *peripheral dark-field* microscopy as commonly used, the illumination is a symmetrical hollow cone of light with an 'internal' numerical aperture greater than that of the objective. In both *central dark-field* and *Schlieren* systems the axial, almost coherent direct light is eliminated in the back focal plane of either the objective or the eyepiece, in the case of central dark-field by a round or slit-like central stop, and in Schlieren microscopy by a knife-edge covering half the aperture.

7.1 OBLIQUE ILLUMINATION

Hooke (1665) was probably the first to employ oblique illumination to improve the visibility of microscopic objects. He wrote 'I never began to make any draught before by many examinations in several lights, and in several positions to those lights, I had discover'd the true form ... Besides, the transparency of most objects renders them yet more difficult than if they were opacous ...'. Hooke (1672–8, quoted by Rienitz, 1975) also used oblique illumination to study 'Schlieren' (streak defects in glass) and imperfections on the surface of lenses. According to Rienitz the '... pupil of the eye plays here, in principle, the same role as the knife edge in the methods of Foucault and Töpler ...' in cutting off the direct light (see the discussion of Schlieren microscopy, below).

Huygens (1703; quoted by Witting, 1906) also recommended oblique illumination to examine Schlieren, and a reflection method to recognize surface defects. Witting however states (contrary to Raveau, 1902 and Rienitz, 1975) that Huygens' methods did not closely resemble those of either Foucault or Töpler.

Baker (1785) described a microscopic method which may also be oblique illumination. He used '... a Cone of black Ivory ...' fastened underneath the object '... principally, when ... the Object very transparent: for Experience teaches, that such Objects are rendered more distinctly visible, by intercepting some Part of the oblique rays reflected from the concave Looking-glass ...'. Dark objects '... will be seen best in a full and strong Light; but if very transparent, the Light should be proportionably weak: for which Reason there is a Contrivance, both in the Single and Double Microscope, to cut off Abundance of its Rays, when such transparent Objects are examined by the greatest magnifiers.'

Dark-field microscopy using oblique illumination was described in a much-quoted article by Reade (1837), who is often credited with having invented the method. 'This new method ... consists in obtaining *oblique refracted light* ...' sometimes combined with slight tilting of the microscope tube relative to the specimen, i.e. '... *oblique vision* as well as *oblique illumination* ...'. The intention here may have been at least partly to obtain improved resolution. Quekett (1852) recommended very oblique light for the '... perfect definition of the markings of certain animalcules of the genus *Navicula* ...', but also stated that slight '... obliquity of the illumination subdues the glare attendant upon perfectly central and full illumination by lamp-light ...'. Using oblique illumination from a right-angled prism, immersed in contact with the slide, Woodward (1878) was able to resolve *Amphipleura pellucida*.

According to Frey (1872) oblique illumination sometimes gives '... truly diabolical illumination ... which however, shows many fine details in an astonishing manner ...'.

More recently Zselyonka and Kiss (1961) and Benedek (1965) have used oblique illumination from a specially constructed condenser to obtain 'plastic'-appearing images, and Kachar (1985) and Sawyer *et al.* (1985) have used oblique illumination with video enhancement to obtain high-resolution pictures of difficult objects. Bretschneider and Teunis (1994) have studied oblique illumination (called by them 'reduced-carrier single-sideband microscopy') both theoretically and practically.

Zernike (1942) comments that since with oblique dark ground there is no difference from which side the spectra are intercepted, one may as well use an ordinary (symmetrical) dark-ground condenser. This is not quite true, and the results of unilateral dark-field illumination are sometimes quite different from those of ordinary dark-ground. The images produced by unilateral dark-field illumination are asymmetrical in the sense that edges perpendicular to the direction of incidence of the illumination are clearly visible while those more or less parallel to the illumination are not. Edges perpendicular to the illumination however tend to appear similar irrespective of whether they lie on the 'near' or 'far' side of the object, and in this respect the images are more symmetrical than those of differential interference contrast or the Schlieren method (see below). Even this generalisation is not quite accurate; with unilateral dark-field illumination the 'near' and 'far' edges of a straight-sided, flat object such as a flat crystal appear similar to each other, but the two edges of a curved or round object do not – this is a simple way to distinguish between these two classes of object.

Oblique dark-field illumination should be distinguished from oblique bright-field illumination in which the direct light just enters the objective. Both variants can be used to study pale or transparent specimens, but as discussed in Chapter 5 the latter can improve the resolving power of the system especially with periodic objects, and tends to give asymmetric images closely resembling those of differential interference contrast.

7.2 SYMMETRICAL PERIPHERAL DARK-FIELD

Reflecting paraboloid or cardioid condensers of high aperture have been specially manufactured for peripheral dark-field (e.g. Wenham, 1878). They seem however to have little advantage over an ordinary, well-corrected refracting condenser used with a central opaque stop in its front focal plane (e.g. Shadboldt, 1851; Abbe, 1873) and with its numerical aperture increased by oil- or water-immersion. The symmetrical contrast obtained with peripheral dark-field is excellent, and particles smaller than the limit of resolution, such as fat droplets (chylomicrons) in blood plasma, can readily be visualized.

7.3 ULTRAMICROSCOPY

Ultramicroscopy is dark-field microscopy employing a very bright light and with the illumination approximately perpendicular to the optic axis of the microscope. It can demonstrate particles which are orders of magnitude smaller than the limit of resolution. *Visibility* is not however the same as *resolution*; the latter implies that not only the presence but also the shape or size of an object can be distinguished.

7.4 RHEINBERG ILLUMINATION

Rheinberg illumination (Abramowitz, 1983; Taylor, 1984) is peripheral dark-field microscopy in which the central part of the front focal plane of the condenser is a coloured filter, not opaque as in an ordinary dark-field condenser, and the outer part of the condenser stop is a filter of a different colour. Edges and discontinuities of the specimen appear in the colour of the outer filter and the background in the colour of the central portion. The images are often very striking and beautiful, but it is not certain that they convey much more information than those of ordinary dark-field.

7.5 CENTRAL DARK-FIELD

Central dark-field has been discovered independently many times (e.g. Ross, before 1855; Töpler, 1882; Exner, 1885; Gordon, 1906, 1907; Nelson,

1910; Siedentopf, 1915; Spierer, 1926; Volkmann, 1928). The method may even have been used by Lister (1830), who wrote 'It is the marginal rays which contribute especially to render visible close and delicate lines, such as those on the scales of lepidopterous insects, and some of the most difficult of these are even best seen when the central light is intercepted . . .'.

Central dark-field may be regarded as an extreme form of phase contrast in which the transmittance of direct light by the phase plate tends to zero, giving symmetrical images of high contrast. According to Zernike (1935, 1942) the method gives a much better representation of the object than does oblique dark ground. In theory it is immaterial whether the central opaque stop is in the back focal plane of the objective or of the eyepiece, but in practice the former position is often inaccessible with high-power objectives, and the latter is usually somewhat inconvenient since the eyepoint is very small, especially with objectives of low numerical aperture.

Just as in the case of 'non-ideal' phase contrast, the central stop in general obstructs not only the direct light but also some lower orders of diffracted light. This results in an effect analogous to the phase 'halo'. Spitta (1909) severely criticized Gordon's central dark-field system, and stated that the absence of the direct light could lead to gross errors in the image. It is however not obvious why such errors should be any more

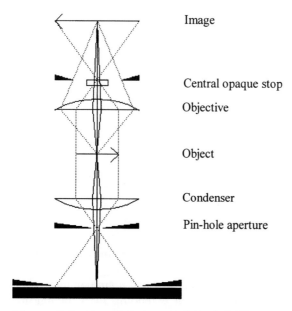

Image

Central opaque stop

Objective

Object

Condenser

Pin-hole aperture

Figure 7.1   Central dark-ground (dark-field)

likely with central dark-field than with conventional (symmetrical) or with unilateral oblique dark-field.

Under some circumstances the edge of an absorbing object appears with central dark-field as a pair of bright lines separated by a dark minimum. This is said to be of zero intensity, and to lie at or very near the position of the geometrical edge (Zernike, 1935, 1942; Birch, 1968; Young, 1983). Siedentopf (1909) regarded this as a disadvantage since it may result in spurious resolution, e.g. of fine grating objects, but the phenomenon can be useful in metrology. For this purpose it has been recommended that the diameter of the central stop should be one quarter that of the objective (Young, 1989). The computer program suggests that provided the diameter of the central dark-field stop is not too small and the objective NA is less than about 0.78, with an object (either absorbing or phase) exactly in focus, a dark but not necessarily zero intensity line appears in the image between two bright lines at or near the geometrical edge. A dip is however also seen even with larger objective apertures (and a dark-field stop of diameter 0.25), provided the object is thrown somewhat out of focus. If these results are correct, it may be supposed that in practice a dip in intensity between two bright lines would almost always be seen at or near the edge of an absorbing object of finite thickness, since even at the 'position of best focus' part of the object would probably be outside the finite depth of focus of the objective.

7.6   SCHLIEREN MICROSCOPY

Töpler (1864, 1866, 1882, 1906) described imaging with a movable opaque slider ('Schieber') in the back focal plane of the objective or eyepiece. He preferred the former site as the eye tends to get in the way if the eyepoint is used, but in 1885 described a commercial attachment consisting of a frame placed above the eyepiece to achieve the Schlieren effect with higher-power objectives.

Test objects such as potato starch grains in balsam were excellently seen in apparent relief (shadowed) when the slide was slowly pushed in, while the background gradually became darker. All grains in the field were affected uniformly, which is never attainable with oblique illumination. Töpler was unable to account for the fact that resolution of fine lines (e.g. of *Navicula angulata*) was lost if the lines lay parallel to the edge of the slider, while lines at right angles appeared surprisingly clearly; the explanation is rather obvious on the basis of the diffraction theory already discussed.

Independently of Töpler, Foucault (1858) devised a similar method to test lenses but apparently did not think of extending the principle for more general scientific purposes (Witting, 1906).

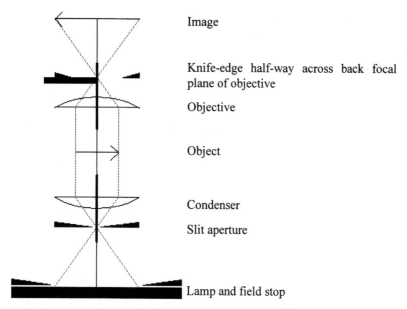

Image

Knife-edge half-way across back focal plane of objective

Objective

Object

Condenser

Slit aperture

Lamp and field stop

Figure 7.2   Schlieren microscopy

Exner (1885) completed his work on the Schlieren method before learning of Töpler's previous articles, and applied the method success-fully to blood cells, living muscle fibres, cells in insect eye, bone and cartilage. He thought that the use of the apparatus to give better microscopic contrast relief might be limited by difficulties in interpreting the striking images, but that its use in immersion refractometry (see below), to establish the matching of the refractive indices of the object and mounting medium, was not liable to the same objection.

7.7 IMMERSION REFRACTOMETRY AND DISPERSION STAINING

The refractive index of microscopic objects has long been studied by immersing the specimen in a series of media of known refractive index, and finding the medium in which it is invisible (see the discussion on the Becke line, Section 3.12.2). The imaging methods used for the purpose in the 19th century included axial illumination, oblique illumination and Schlieren microscopy (Töpler, 1864; Exner, 1885). Exner studed living cells and tissues in aqueous media, and determined refractive indices significant to the third decimal place in media ranging from olive oil (RI 1.4730) to cadmium borotungstate (RI 1.6066).

As an alternative to using a series of media with light of a given mean wavelength, a single medium of high dispersion has sometimes been used with quasi-monochromatic light of varying wavelength.

Maschke (1872) used axial white light illumination to distinguish between microscopic particles of quartz and tridymit, immersed in a medium with a refractive index intermediate between those of the two substances. At a particular focus of the microscope particles of quartz appeared blue or blue-green with a reddish edge, while tridymit appeared reddish with a blue or blue-green edge.

Schroeder van der Kolk (1892, 1898) used oblique illumination for the microscopic study of objects immersed in liquid media of high dispersion, and Wright (1913) described several alternative methods to obtain suitable oblique illumination. Wright wrote that 'In case the mineral grain and liquid have the same refractive index for some wave length near the centre of the visible spectrum (about $550\mu\mu$), the greater color dispersion of most liquids over solids enters the problem and gives rise to characteristic color fringes along the margins of the mineral grain . . .'. The effect could be seen with a narrow central cone of light from the condenser, but was clearer with light from a strong point source ('Nernst glower'), so oblique as effectively to give unilateral dark-field illumination.

Crossmon (1948) used dark-field illumination and mounting media of high dispersion to study unstained tissue by 'optical staining'. He acknowledged the relationship of the technique to the oblique illumination method described by Wright (1913) for petrographic work, and in 1949 introduced the term 'dispersion staining'. Dispersion staining is currently popular for the routine study of such objects as dust particles and asbestos fibres (Brown and McCrone, 1963). If correctly used, different materials appear with characteristic and distinctive colours and are readily recognised by an experienced worker.

Chapter 8

Gegenfeld

8.1 PRINCIPLE OF GEGENFELD

In this relatively little-known method a half-wave retarding plate with a straight edge is placed over one half of the back focal plane of the objective so that it affects the diffracted light on one side only of the central, direct light. Striking images, especially of the edges of transparent objects, can be obtained.

The first description of the effect of reversing the phase of half the diffracted light appears to have been by Bratuscheck (1892), who used either a glass wedge, or a quartz biplate which rotated the plane of polarisation of each diffracted beam through 90° in opposite directions. He noted that the apparent position of bars and slits in a grating object were exchanged. Subsequently Zernike (1942) compared phase contrast with the Schlieren method, oblique dark ground, central dark ground, and a narrow central pencil. 'Instead of by intercepting the diffracted light on one side, the relief effect may also be brought about *by reversing the phases on one side* ... This improved form of the Schlieren method therefore doubles the sensitivity ...'. The '... original Schlieren method does not materially alter the image of an amplitude grating, while the improved form will, to the same approximation, make it invisible ...'. Wolter (1950a) claims to have invented the 'Gegenfeld' method: 'Die Bedeutung des Gegenfeldverfahrens, das Verfasser als Schlierenverfahren einführte, tritt erst bei komplizierteren Objekten hervor'. ("The significance of the Gegenfeld method, which the author introduced as a Schlieren method, is seen with more complicated objects"). To a fuller

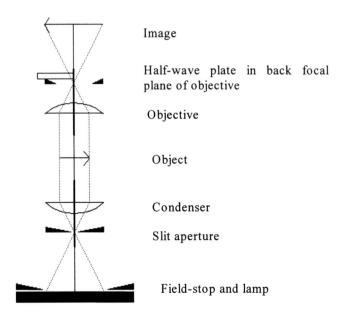

Image

Half-wave plate in back focal
plane of objective

Objective

Object

Condenser

Slit aperture

Field-stop and lamp

Figure 8.1   Gegenfeld

description of the method Wolter (1950c) added a note that after the
investigation was complete he came across a reference to Zernike (1934),
suggesting the use of a half-wave plate. He wrote that Zernike erro-
neously thought the method gives the same results as a knife-edge, but
still seemed unaware of both Zernike's 1942 article and the yet earlier
work of Bratuscheck.

Various methods which are more or less equivalent to the use of a half-
wave phase plate over one side of the Fourier plane have been described
several times. Misell *et al.* (1974) proposed that in bright-field optics '...
the method of recording complementary half-plane images can be used to
determine the amplitude and phase of the object wavefunction ...'.
Dekkers and de Lang (1974) showed that if the photodetector in a
scanning electron microscope is divided into two semicircles, subtracting
the outputs gives 'differential phase contrast' with increased image detail
and edge enhancement. They suggested that the method might also be
applied in optical scanning microscopy. In 'point-source' scanning optical
microscopy, according to Hamilton and Wilson (1984) and Sheppard
(1987), the difference of the outputs of the two halves of a detector is
directly proportional to weak phase gradients in the specimen. The
method has been used in various instruments (see e.g. Horikawa *et al.*,
1987).

The image obtained with 'Gegenfeld' depends critically on what
happens to the zero-order light. According to Wolter the amplitude of

the non-diffracted light is zero and the symmetrical image thus obtained is a form of dark-field. This could in practice occur if the edge of the half-wave plate in 'Gegenfeld', or the junction between the two photo-receptors in scanning microscopy blocks the direct light. If however at least some of the direct light reaches the image plane, for example if the junction of the split detector is slightly to one side, the image is asymmetrical and resembles that found with differential interference contrast. This is the method described by Zernike. Both types can be simulated in the program, which allows complete, partial or zero transmission of the zero-order light.

Chapter 9

'Modulation Contrast'

9.1 PRINCIPLE OF MODULATION CONTRAST

In this imaging system (Hoffman and Gross, 1975a, 1975b) the illuminating beam is restricted by a narrow slit in the front focal plane of the microscope condenser. with three parallel zones of different absorption.

Situated in the conjugate plane at the back focal plane of the objective is a 'modulator'. The image of the condenser slit falls on a narrow central strip of the modulator, which typically absorbs 85% of the undiffracted light. On one side of the optic axis the diffracted light passes unabsorbed through a clear zone, and on the other side the diffracted light is 99% absorbed by a dark region. The modulator is said not to affect the relative phases of the beams of light passing through the various zones. In practice the central zone is slightly wider than the image of the condenser slit, so that a small part of the diffracted light passes through the central zone together with the direct light. This affects the functioning of the apparatus in a non-ideal way (cf. practical phase contrast microscopy), and is simulated in the program by making the central zone a finite width.

Non-diffracted light transmitted by homogeneous regions of the background or specimen passes entirely through the central modulator zone. The image of such regions is unaffected by the modulator apart from a uniform reduction in intensity. Light diffracted by phase and/or intensity gradients and edges in the specimen however undergoes differential absorption in the modulator, which therefore affects the contrast of the final image. The asymmetry of the optical system tends to cause

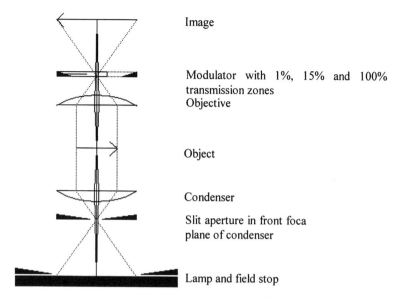

Image

Modulator with 1%, 15% and 100% transmission zones
Objective

Object

Condenser

Slit aperture in front foca plane of condenser

Lamp and field stop

Figure 9.1   Hoffman's 'Modulation Contrast'

asymmetry of the image, with gradients and edges on two opposite sides of the object respectively appearing darker and brighter than the background. The image appears somewhat three-dimensional, and resembles that seen with differential interference contrast (Hoffman and Gross, 1975b).

Because the numerical aperture of the condenser is severely restricted in one azimuth, in normal use the resolving power of the system in a direction perpendicular to the slit is effectively reduced to that found with coherent illumination. The modulator can however be offset, so that (for example) the grey, 85% absorption zone can be situated at the edge of the exit pupil of the objective, and the most highly absorbing region is completely outside the objective aperture and is therefore ineffective. The illuminating slit below the condenser is simultaneously offset to ensure that the direct light continues to pass through the grey zone of the modulator. Under these conditions the net effect may be regarded as a combination of unilateral oblique illumination and apodization, and there may be little or no impairment of the resolving power. Because the modulator is asymmetrical, both the direction and the extent of its movement are important in determining the final image.

The imaging system is extremely flexible, because it is possible not only to alter the position of the modulator but also to exchange one type of modulator for another. Depending on the type and adjustment of the modulator, improvement of image contrast may in theory be expected with some phase objects, amplitude objects, and objects of relatively small dimensions. There is however in general no linear relationship between physical properties of the object and the intensities observable in the image, and just as with phase contrast microscopy reversal of contrast can occur with some specimens. Interpretation of results can therefore be very difficult.

In recent commercial instruments the modulator is fixed in the back focal plane of the objective with the grey (partially transmitting) section almost at one edge of the aperture and the dark (least transmitting) section lateral to this. A slit in the front focal plane of the condenser is adjusted so that its image falls on the grey region of the modulator. Adjacent to this (unalterable) slit is a second aperture in a position conjugate with the completely transparent portion of the modulator; the transmittance of this aperture can be varied by rotating a polarizer situated in front the slit. For maximum image contrast no light is allowed to pass through the variable part of the slit, but contrast can be lowered by increasing the transmittance of the variable part of the slit, if for example there are very large differences in refractive index between the specimen and the surrounding medium.

A fixed modulator can be situated between the elements of a high-powered objective, nearer the back focal plane than is possible with an adjustable or removable modulator. This is however at the cost of some loss of flexibility of the system. The transmission of only 1% of the diffracted light through the narrow region lateral to the partially transmitting zone would have little if any effect on the image, and such a system would probably behave in much the same way in principle as a Schlieren system, or indeed simple oblique coherent illumination. The Hoffman system does however use almost the full objective aperture to catch the diffracted light, instead of only half the aperture as is the case with Schlieren microscopy using axial coherent illumination – the Hoffman system would therefore have substantially better resolution. In addition the contrast in the Hoffman system is increased (as in a typical phase contrast microscope) by absorbing most of the direct light in a controlled and reproducible way, thus making the intensity of the direct light similar to that of the diffracted light and allowing more complete interference. This is an advantage over simple oblique illumination, in which the intensity of the direct light is variably affected by the extent to which it is intercepted by the edge of the condenser aperture.

In the computer model, 85% of the zero-order diffracted light is absorbed by the central zone of the modulator, which is 0.125 NA wide. The diffracted light is 99% absorbed on one side of the central zone and not at all on the other. 'Offsetting' of the modulator to one side or the other of the optic axis is possible, and when the offsetting is specified a reminder of the current obliquity of the coherent illumination is given. If the offsetting equals the obliquity, the situation is equivalent to offsetting the modulator and simultaneously moving the slit below the condenser, as recommended by Hoffman.

Commercial instruments are available from a number of suppliers, including Modulation Optics Inc. (100 Forest Drive at East Hills, Greenvale, N.Y. 11548, U.S.A.), Nikon. and Olympus.

Chapter 10

Apodization

10.1 DEFINITION

Strictly speaking 'apodization' means 'removing the feet', i.e. suppressing the side lobes of the diffraction-limited image of a point. This improves the ability of a system to resolve two nearby objects of markedly unequal intensity, a situation common in astronomy, and observation of very high-contrast detail, especially in microscopy. It can be achieved by reducing the intensity of the diffracted light progressively and gradually towards the edge of the exit pupil, without altering the relative phases (Jacquinot, 1950; Jacquinot and Roizen-Dossier, 1964). The procedure however diminishes the resolving power for objects of equal brightness, and in general apodization '... achieves only an improvement of certain desirable qualities at the expense of others ...' (Jacquinot and Roizen-Dossier, 1964). The technical difficulty of manufacturing an apodization plate which absorbs some light without having any effect on retardation need not concern us here.

The term 'apodization' is applied rather loosely by some authors (several of whom are listed by Jacquinot and Roizen-Dossier, 1964) to *any* modification of the pupil amplitude function, and this definition is followed here. Accordingly, 'apodization' includes 'super-resolution', i.e. an improved ability to resolve two points of equal brightness. This can be obtained by reducing the amplitude transmission function mainly at or near the pupil centre and allowing it to remain unaltered near the edge of the aperture – this decreases the width of the central maximum of the image of a point object but is usually accompanied by a considerable rise

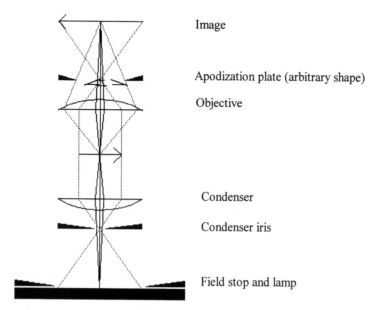

Image

Apodization plate (arbitrary shape)

Objective

Condenser

Condenser iris

Field stop and lamp

Figure 10.1 Apodization

in the level of the 'feet', and is thus in a sense the opposite of true apodization. Because of the associated change in the 'feet' and a reduction in the intensity of the central maximum of the image there is no point in reducing the width of the central maximum excessively. In practice, apodization may in general give a gain in resolution of about 25% or less.

As with many other methods of microscopy, apodization was anticipated by Bratuscheck (1892), who noted that attenuating the direct light gives improved contrast in the images of object details with small absorption or retardation.

In an 'Amplitude Contrast' microscope designed by M. Pluta and manufactured by Polskie Zaklady Optyczne (Warsaw, Poland) the phase ring of a conventional phase contrast microscope is replaced with a ring of soot backed with a thin dielectric layer. This absorbs 75–80% of the direct light with little or no effect on the phase, and is claimed to give improved contrast and sharpness in images of small amplitude details. The relationship to apodization is explicitly recognised in the company's advertising literature.

10.2 SIMULATION OF APODIZATION

The apodization option in the Zernike computer program can simulate almost any amplitude and retardation function in the back focal plane of the objective, and can be used as a somewhat more complicated but equivalent alternative to other imaging options. The user first specifies the maximum absorbance and maximum retardation of the plate, and then its thickness as a function f(X) of the distance X of a given point in the diffraction pattern from the optic axis. The function can be symmetrical on both sides of the optic axis, or variable across the whole back focal plane of the objective.

For example, if f(X) simply equals X, the thickness of the plate increases linearly from the centre of the pupil to the side, and the intensity of the transmitted light therefore decreases logarithmically. Similarly, if the function equals −X the absorbance decreases and the transmittance increases from the optic axis outwards. The functions (X*X and (−X*X) produce curved thickness functions. The retardation and absorbance at a given distance from the optic axis are scaled in proportion to the specified maximum retardation and absorbance. Logical (Boolean) functions can be used alone or in combination with algebraic expressions. Thus the function (X > 0.25 AND X < 0.75) implies an effect only between NA 0.25 and NA 0.75, and (X*X)*(X > 0.5) produces a curved thickness function above NA 0.5 with no effect below this value. Note that the BASIC multiplication sign (*) between the two halves of the function indicates that they act simultaneously. Remember also that a Boolean expression evaluates to 0 if it is untrue and −1 if true; the whole expression (X*X)*(X > 0.5) therefore is −(X*X) if X is greater than 0.5 and zero if it is not. In the rather complicated function (X*X)*(X > 0.5) + (X)*(X = <0.5) the addition sign indicates that the left side of the function acts over a particular range of values, and the right side over another.

Irrespective of the screen display mode chosen for the diffraction pattern (intensity, amplitude, etc.), the program demonstrates the effect of the specified apodization plate by superimposing on the display the intensity and retardation due to the plate. This enables various apodization functions to be experimented with.

The effects obtainable by apodization depend in a complicated way on the chosen function, and include improving the relative contrast and visibility of small and/or slightly absorbing objects. As the transmittance of a central stop decreases to zero, apodization eventually becomes indistinguishable from central dark-field.

10.2.1 Computed effect of apodization on resolution. The following table illustrates the improvement in resolution (defined on the Sparrow criterion) predicted by the Zernike program using apodization, coherent illumination, a field width of 64 half-wavelengths, and two bright point objects one wavelength apart.

NA	Maximum absorbance of the apodization plate	Maximum retardation of the apodization plate	Apodization function f(X)
0.6562	1.6	0	COS(X*π/2)
0.625	3	0	COS(X*π./2)
0.6406	2.0	0	$1 - X*X$
0.625	2.5	0	$1 - X*X$
0.625	2.5	0	$1 - X$
0.625	2.5	0	$NA - X$ (equiv. to $0.625 - X$)
0.5312	3.0	0	$x < 0.5$

Of the limited selection of apodization plates tested the last (a simple 'top-hat' shape which absorbs most of the central light) seemed to give the best results. Under similar conditions, the minimum objective NA required for resolution without apodization was 0.6875 using coherent illumination and 0.4375 using incoherent illumination. With coherent light apodization therefore gave a significant improvement in resolution; but whether a similar improvement might be obtained with incoherent or partially coherent illumination remains to be investigated.

Chapter 11

Fluorescence

Fluorescence is a phenomenon in which the energy of light (the *exciting radiation*) striking an object raises electrons of the absorbing molecules to higher orbits. On falling back to their original orbits, the electrons emit some or all of the excess energy as light, which according to Stoke's law in general has a longer wavelength than the exciting radiation. The longer the wavelength of light, the less energy there is per quantum; the 'missing' energy is dissipated in the form of heat, or converted into chemical energy which can damage the absorbing molecules and cause them to fade.

The fluorescent material in microscopic material may be an intrinsic part of the specimen, such as Vitamin A, or may be a fluorescent dye attached more or less specifically to the specimen, perhaps via an antigen–antibody link. In recent years numerous highly specific and useful fluorescent staining techniques have become available. Using these it is possible to demonstrate various intracellular organelles and to detect and measure such things as intracellular pH, calcium and other substances, and membrane-bound sites of various kinds, and with suitable apparatus it is possible to study extremely rapid intracellular processes. It is largely due to these developments, which are beyond the scope of the present book, that fluorescence microscopy, especially confocal scanning fluorescence microscopy (see Chapter 12), has become so popular.

The colour (band of wavelengths) of the emitted light depends mainly on the energy differences between the electron orbits concerned, and is partly or wholly independent of the wavelength of the exciting light. The *number* of quanta emitted, and hence the intensity of the fluorescence, is however related to the wavelength of the exciting radiation, and to how efficiently this is absorbed by the object. The probability of a quantum of exciting light of a given wavelength being absorbed by a molecule is a function of the colour of the molecule, and only energy which is absorbed can excite fluorescence.

11.1 EXCITING AND BARRIER FILTERS

The intensity of the emitted light is usually much less than that of the exciting radiation. Scattered or stray exciting light reaching the detector (eye, camera or photo-multiplier tube) would therefore completely swamp the emitted light unless this is prevented by some optical means. Commonly a pair of matched colour filters is used – an *exciting filter* in the illuminating beam passes only light of relatively short wavelength, and a *barrier filter* between the object and the detector passes only the longer wavelengths of the emitted fluorescent light, and cuts out most or all of the light passed by the exciting filter. With perfectly 'crossed' filters no exciting light would reach the detector, but this ideal is not always attained – although it is quite easy to obtain a 'well-crossed' pair of filters if the spectra (wavelengths) of the exciting and the emitted light are very different, in practice the spectra often overlap to a considerable extent. Thus the fluorescence of the familiar dye eosin can be excited quite well by ultra-violet or short-wave blue light, but is much better excited by visible green light with a wavelength close to that of the emitted radiation. Green light is therefore considerably more efficient than ultra-violet light in exciting eosin fluorescence. Since the wavelength of the light emitted by eosin is only slightly longer than that of the best-absorbed exciting light, it is impractical to use crossed colour filters with eosin fluorescence excited by green light; instead, crossed polarizing filters have been used in conjunction with dark-field microscopy (Goldstein, 1969).

If the exciting light is entirely in the (invisible) ultra-violet region, or is very short-wave violet light, the barrier filter can be pale yellow or almost colourless, and will scarcely affect the apparent colour of the emitted fluorescence. When using blue or longer-wavelength exciting light, however, the barrier filter is necessarily coloured, and can partially absorb some of the emitted light and thereby markedly affect its perceived colour.

11.2 TRANSMITTED VS. INCIDENT ILLUMINATION IN FLUORESCENCE MICROSCOPY

Until a few decades ago, fluorescence microscopes mostly utilized conventional transmitted illumination. Sometimes an ordinary bright-field condenser was used together with crossed coloured exciting and barrier filters to prevent the exciting light reaching the detector. Some workers however preferred a dark-field condenser which illuminated the object with a hollow cone of light of high numerical aperture, none of which could directly enter the objective (of relatively low aperture); in this case the demands on the exciting and barrier filters were less severe. More recently it has become usual to utilize incident illumination (Figure 11.1) through the objective itself, the necessary separation of the incident and fluorescent beams being attained by the use of a beam-splitter. Crossed filters are still necessary to cut off exciting light which is reflected off the back of the objective, the coverslip and the microscope slide.

Incident fluorescence illumination has a number of advantages. For example, once the illumination has been correctly adjusted, changing the focus of the image by changing the relative position of the objective and object automatically changes also the focus of the illumination, so that the illuminating field stop remains sharply in focus in the field of view. In addition, if the object is very strongly absorbing, much of the emitted fluorescent light will be reabsorbed by the specimen when using a transmitted-light system, while with incident illumination light emitted from the upper layers of even an effectively opaque specimen is able to reach the detector. In practice, however, conventional transmitted-light illumination works reasonably well.

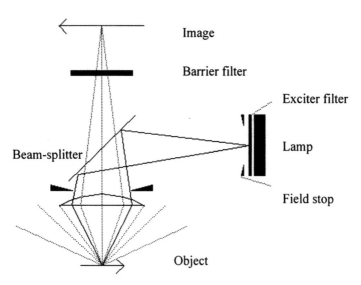

Figure 11.1   Incident-light fluorescence

11.3   LOSS OF LIGHT AT AN INCIDENT-LIGHT BEAM-SPLITTER

Assuming the beam-splitter in an incident-light fluorescent system transmits half the light striking it and reflects the other half, it will readily be seen that three-quarters of the possible intensity of the visible image will be lost – half of the illuminating beam is lost, and half of the emitted light. A 50:50 reflection:transmission ratio can only be attained with a dielectric (non-metallic) device relying on the partial reflection of the incident light by thin layers of material of different refractive index; a metallic beam-splitter such as a half-silvered mirror necessarily absorbs some of the light and cannot therefore be as efficient. Dichroic beam-splitters which reflect perhaps 70% of the incident, short-wave radiation and transmit 70% of the emitted long-wavelength radiation are clearly more efficient when using the particular wavelengths for which they are designed, giving an overall efficiency of almost 0.5. They however have the disadvantage of performing much worse with other combinations of exciting and emitted wavelengths, and they can also to some extent affect the perceived colour of the fluorescent image, behaving in this respect like a coloured barrier filter. Commercial incident-light fluorescent microscopes are commonly equipped with dichroic beam-splitters, sometimes mounted in cubes which also incorporate a particular exciting filter and barrier filter. Such devices are suitable for routine work with a particular fluorescent dye, but for more varied work a fixed neutral beam-splitter used with sets of interchangeable exciting and barrier filters is often more convenient.

11.4   COMPARISON OF TRANSMITTED AND INCIDENT-LIGHT FLUORESCENCE ILLUMINATION WITH LOW- AND HIGH-APERTURE OBJECTIVES

As already pointed out, the intensity of the image is in a real microscope directly proportional to the square of the numerical aperture of the objective. The intensity of the fluorescence is also (approximately) proportional to the intensity of the illumination, which is directly proportional to the square of the aperture of the condenser. In an incident-light system the condenser and the objective are in fact the same lens, so that the final image intensity is directly proportional to the fourth power of the lens aperture. This is ideal when using an oil-immersion objective of high aperture, the results then being identical with that obtained with transmitted illumination with an oil-immersion condenser of aperture equal to that of the objective (ignoring possible difference in the exciting and barrier filters needed, and losses in the incident-light beam-splitter). With low-power objectives, however, the intensity drops off very rapidly with incident-light illumination, while with transmitted illumination it is possible to use a high-aperture condenser even with a small-aperture objective. The advantage of doing this is clearly seen in the following table.

Table 11-1

Condenser aperture	Objective aperture	Relative intensity
1.3	1.3	2.8561
0.25	0.25	0.0039
1.3	0.25	0.106

For good results with either a transmitted-light or an incident-light fluorescence microscope it is essential for the illuminating system to incorporate a suitable field-stop, an image of which is sharply focussed onto the specimen. This eliminates possible fading of parts of the specimen not currently being observed, and prevents stray fluorescent light from such parts entering the objective and degrading the image.

11.5   FULL-APERTURE VS. DARK-FIELD ILLUMINATION

With both transmitted-light and incident-light fluorescence, it is possible to use the full aperture of the illuminating lens. Alternatively, one may use only the outer part of the aperture, so that the hollow cone of exciting radiation cannot directly enter the objective (or the imaging part of the objective aperture in incident-light illumination). Dark-field illumination requires somewhat less stringent crossing of the exciting and barrier filters, but in practice this advantage is usually minimal. A serious disadvantage of the method is the fact that the intensity of the illumination is greatly reduced by the reduction in the effective illuminating aperture. Assume that the (transmitted-light) condenser has an immersed numerical aperture of 1.4, which is about the maximum obtainable. If the inner aperture of the hollow cone of light is about 1.0, permitting the use of an objective of NA 0.95, the intensity of the illumination is proportional to $1.4^2 - 1.0^2$, i.e. is proportional to 0.96. If the full aperture of the condenser is used the intensity of illumination is proportional to 1.96, just over twice as great. Usually, the advantage of using dark-field illumination in reducing the need for heavy barrier filtering is more than outweighed by the unavoidable loss of illuminating intensity. In addition, with dark-field transmitted illumination the aperture of the objective is necessarily limited, further diminishing the intensity of the final image.

11.6 FADING OF FLUORESCENCE

As already indicated, degradation of the fluorescent material by short-wave radiation can be a serious problem in quantitative fluorescence microscopy. It may be necessary to find the specimen using dim visible light and perhaps phase-contrast illumination, and then study or photograph the fluorescence using as short an exposure as possible. The problem is to a large extent eliminated with confocal microscopy, in which individual points in the object are only fleetingly exposed to the exciting radiation.

11.7 QUANTITATIVE FLUORESCENCE MICROSCOPY

It is often assumed in fluorescence microscopy that the intensity of fluorescence is directly proportional to the concentration of the fluorescing material. This assumption is seldom justified. In a test-tube, the intensity of the emitted fluorescence of an aqueous solution of a dye is frequently found to be linearly related to the concentration only with very low concentrations. With higher concentration *autoquenching* occurs; reduction in the emitted fluorescence intensity is caused by interaction between molecules of dye, which results in energy being dissipated as heat. The phenomenon is typically much reduced in less polar solvents such as xylene or ethanol, and the influence of the surrounding mounting medium can be marked in microscopic preparations. Quenching can also be caused by traces of heavy metals or elements such as iodine in the medium or preparation.

Strict proportionality between the intensity of fluorescence and the concentration of the fluorescent molecules is not to be expected even theoretically. According to the Bouguer–Beer law (see Section 13.1), if a parallel beam of monochromatic light passes through a homogeneous medium the absorbance is

$$D = a.b.c \qquad (11.1)$$

where a is the absorptivity, b is the path-length through the solution in some suitable unit, and c is the concentration. The absorbance is defined as

$$D = -\log\left(\frac{I_t}{I_0}\right) = -\log(T) \qquad (11.2)$$

where I_0 is the intensity of the original radiation, I_t is the intensity of the light leaving the solution, and T is the transmittance. Under ideal circumstances the intensity of the emitted fluorescent radiation is proportional to $1 - T$, which is approximately proportional to the absorbance (and hence to the path length and to the concentration) only when these are small. In practice, the concentration of fluorescent material in microscopic preparations is often very low, and the lack of proportionality between concentration and fluorescence may sometimes be ignored.

Fluorescence ratio imaging has recently become popular for the measurement ratio imaging of materials such as Ca^{++} or hydrogen ions in living cells. The method depends on the use of dyes, the colour of whose fluorescence is a function of the concentration of the ions to be measured. The *ratio* of the intensities at two different wavelengths is measured, and compared with calibration curves. The measurements are to a large extent independent of both the concentration of dye, the intensity of exciting radiation, and the absolute sensitivity of the detector.

In traditional fluorescence microscopy the object is excited with relatively short-wave light, and according to Stoke's law the emitted photon is in general of lower energy and hence longer wavelength. In *two-photon fluorescence* (Denk *et al.* 1990) the target molecule is struck simultaneously, or at least in very quick succession, by two photons of red or infra-red light, the combined energy of which is sufficient to allow *anti-Stokes radiation*, i.e. the emission of a single shorter-wave photon. Because two simultaneous exciting events are required, the probability of a photon being emitted is proportional not to the intensity but to the square of the intensity of the exciting radiation, and hence to the fourth power of the numerical aperture of the illuminating lens. The extremely high intensity required can be provided, with some technical difficulty and at considerable expense, by a laser which is able to produce many million intense, very short pulses per second. Because the pulses are so short (of the order of a few femtoseconds) the time-averaged power of the laser can be quite low, perhaps 1W for a laser producing pulses with a power as high as 10 kW (Cannell and Soeller, 1997).

Using an incident-light optical system, the intensity of illumination is at its maximum at the focal point of the objective, while in other places it is approximately inversely proportional to the square of the distance from this point. The intensity of two-photon fluorescence at a given point in space is therefore approximately inversely proportional to the fourth power of the distance of the point from the focal point of the objective, i.e. it drops off very rapidly indeed away from the plane of best focus. This enables two-photon fluorescence to give excellent 'optical sectioning' of thick objects, comparable with that obtained with confocal scanning (see Chapter 12), even without a pin-hole aperture on the imaging side of the system and despite the fact that the wavelength of the exciting light in two-photon fluorescence is relatively long. By using two-photon fluorescence in combination with confocal scanning even better optical sectioning can sometimes be obtained, and this approach will probably be widely used if commercial instruments become available at a reasonable price.

Two-photon fluorescence microscopes can often work at even greater depths within a thick specimen than can confocal scanning instruments, due to some extent to the fact that red light penetrates better than does light of shorter wavelength, and partly due to the fact that the resolution is not affected by scattering of the emitted radiation.

Tissue damage often results from the intense radiation in two-photon fluorescence, but is usually limited to the immediate region of the focus, a volume of perhaps one cubic micrometre. This is much smaller than the region damaged in conventional fluorescence microscopy (Dixon, 1997).

11.8 COMPUTER SIMULATION OF FLUORESCENCE

The computer model takes account of (or assumes!) the following:

(1) Fluorescent light emitted by different molecules is incoherent. The resolution using an objective of given aperture is therefore equivalent to that found with an 'ideal' bright-field condenser having an illuminating half-angle of 90° (i.e. with the NA of the 'dry' condenser, unattainable in practice, equal to 1.0).

(2) In the one-dimensional computer model, the intensity of the image is directly proportional to the numerical aperture of the objective. In reality, with a two-dimensional circular objective aperture, the visible intensity would be proportional to the square of the objective aperture.

(3) The intensity of the emitted radiation is directly proportional to the intensity of the light absorbed, i.e. equals $1 - T$ where the original intensity of the radiation striking the object is unity and T is the transmittance of the object.

The computer model however ignores many factors which would substantially affect the actual intensity of a fluorescent microscopic image, including self-absorption of the emitted light, fading of fluorescence, polarization of fluorescence (quite common with oriented dye molecules) and losses in the optical system.

Chapter 12

Confocal Scanning

12.1 INTRODUCTION

Confocal scanning microscopy is a relatively new technique already widely used in both biology and material science (see McMullan, 1990, for its early history). The present chapter, based mainly on the books of Sheppard (1987), Hegedus (1990), Wilson (1990) and Matsumoto (1993), is an elementary introduction to a increasingly important subject, the theory, apparatus and possible applications of which are currently developing very rapidly.

In a 'Type 1' scanning microscope (Sheppard and Choudhury, 1977), which can in principle use either incident or transmitted illumination, a pin-hole is present in an optical plane conjugate with the specimen, in either the illuminating or the imaging part of the system but not both. Some form of mechanical or electronic device moves the pin-hole and the object relative to each other in a regular two-dimensional scan pattern (raster), and the light transmitted, reflected or emitted by the object is collected by a lens and passed to a detector such as a photomultiplier tube. During the scan a complete image is built up and displayed on a cathode-ray tube or computer screen. The collected data are relatively easy to quantify and analyse digitally, but the image is essentially equivalent optically to that obtained. using conventional microscopy with partially coherent illumination (Welford, 1972).

The optical situation is quite different in confocal ('Type 2') scanning microscopes. Here the object is, as before, illuminated with a small spot of light which moves relative to the object. Light from the object is however focussed by the objective lens onto a *second* pin-hole, which is ideally smaller than the diffraction-limited image of the object point (Airy disc). Only the central part of the Airy disc is therefore allowed through to the photodetector, resulting in alterations to the imaging process including an improvement in both the lateral resolution and the depth discrimination of the system.

12.2 PRACTICAL IMPLEMENTATION OF CONFOCAL SCANNING

Quartz-iodine incandescent lamps and mercury arcs have been used in some instruments, especially of the direct-view type, but in most commercial confocal scanning microscopes a laser produces an intense spot of light at one or more defined wavelengths. The argon ion laser is the most common type in current use, and produces several wavelengths in the middle of the visual spectrum. Lasers which produce ultra-violet light are obtainable, but tend to be expensive and relatively troublesome.

Relative movement of the object and the pin-holes can be achieved in several ways, each with its own advantages and disadvantages.

12.2.1 Object-moving systems. Both pin-holes are stationary and the object itself is moved, generally on a computer-controlled motorized stage (Figure 12.1). Image data are captured on a computer which assembles and processes the final image.

In systems of this type stable alignment of the pin-holes is fairly easy, and very large objects can in principle be scanned. Since only the centre of the microscopic field is ever used, off-axis aberrations such as coma and distortion can be ignored, considerably simplifying the design and construction of lenses. Scanning however tends to be relatively slow, a raster typically taking several seconds even with a fairly small object.

12.2.2 Beam-scanning systems. With the specimen stationary, a reduced image of the illuminating spot is produced in the object plane by a lens which in an incident-light system serves as both condenser and objective. In some systems the light source is a flying spot on the face of a cathode-ray tube, but more typically the relative movement of the object and the illuminating spot results from vibrating mirrors between the condenser and a stationary, illuminated pin-hole. One mirror is responsible for rapid movement of the spot to and fro in (say) an east-west direction, while a second mirror simultaneously moves the spot more slowly from north to south, so that a rectangular area of the specimen is covered.

The incident-light confocal beam-scanning system illustrated (Figure 12.2) is grossly simplified. Not only are the scanning mirrors omitted, but

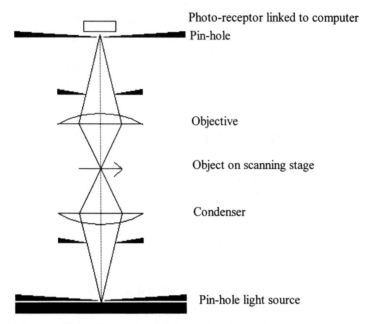

Photo-receptor linked to computer

Pin-hole

Objective

Object on scanning stage

Condenser

Pin-hole light source

Figure 12.1   Transmitted-light confocal scanning

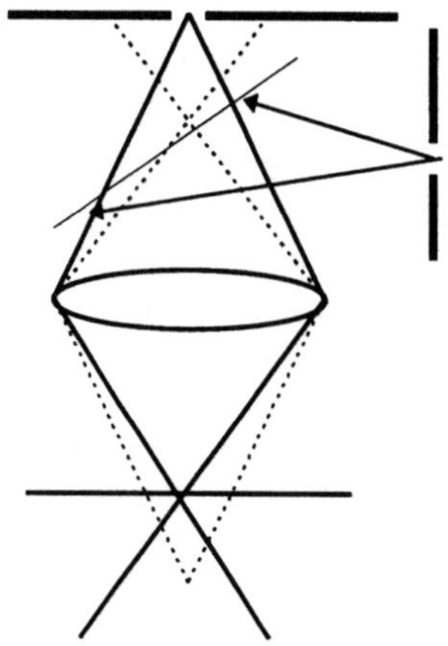

Measuring pin-hole

Illuminating pin-hole

Lens

In-focus object

Out-of-focus object

Figure 12.2   Incident-light confocal system

in reality additional optical components are required to ensure that the lens apertures are fully filled and 'vignetting' is avoided (Wilson 1990b).

12.2.3   Tandem scanning systems.   Tandem scanning, also known as direct-view scanning (Wilson 1990b), uses a device originally developed by Nipkow for early television experiments. A rapidly spinning disc contains a large number of small holes, generally arranged in a spiral, and is illuminated with either monochromatic or white light in a plane conjugate with the specimen. All parts of the object are therefore illuminated at some time during each revolution. On the imaging side of the system a second, identical disc revolves synchronously with the first, so that the diffraction-limited image (Airy disc) of a given illuminated point of the object falls precisely on a corresponding pin-hole. In order to avoid 'cross-talk' between adjacent pin-holes, i.e. light from a given object point falling on more than one pin-hole, the pin-holes must be widely separated, by a distance which depends on both the numerical aperture and the magnification of the lenses.

Provided the discs rotate at more than about 30 revolutions per second the image can be photographed by a simple camera, or viewed directly by the eye without visible flickering. By using a white light source expensive lasers can be dispensed with, and the image is relatively easy to interpret since it resembles the familiar one obtainable with an ordinary microscope.

A disadvantage of tandem scanning, especially in fluorescence microscopy, is that due to the necessarily wide separation of the pin-holes each object point is only illuminated for perhaps 1% of the time, so that the time-averaged intensity of illumination tends to be low. Particularly in transmitted-light systems it is rather difficult to arrange perfect matching and synchronicity of the illuminating and imaging discs, and this difficulty is minimized in some incident-light reflectance and/or fluorescent instruments which contain only a single Nipkow disc; at a given moment a pin-hole on one side of the disc serves as illumination, while a matching pin-hole on the opposite side of the disc receives the image.

12.3 ADVANTAGES AND SOME LIMITATIONS OF CONFOCAL SCANNING

In transmitted-light confocal scanning using small spot sizes, the light taking part in image formation is essentially coherent. In fluorescence confocal scanning, on the other hand, the emitted light is completely incoherent, and in incident-light microscopy light may be reflected either specularly or diffusely. Since image formation is different in all these cases, quantitative and even qualitative conclusions applicable to one set of instrumental conditions may not hold for another.

12.3.1 Reduction of glare (stray light). In conventional microscopy stray light (Section 1.8), mainly due to multiple reflections at glass–air surfaces in the objective, can significantly degrade the image. Stray light can be controlled by limiting as far as possible the area of specimen which is illuminated, and is effectively eliminated in confocal scanning microscopy because the field illuminated at a given moment is so small. As a result confocal scanning images generally have excellent contrast and appear strikingly sharp.

12.3.2 Lateral resolving power. Lateral resolution in microscopy can be characterized (Martin, 1966, p. 230) by a factor $K = xN/\lambda$, where x is the distance apart of two objects, or the interval of a grating, just resolved using an objective of numerical aperture N and light of wavelength λ. The value of K depends not only on the optical system of the microscope, but also on the nature of the test object and the criterion (e.g. Rayleigh or Sparrow – see Section 3.9) adopted in the definition of resolution.

In confocal scanning the lateral resolving power is in general somewhat better than in conventional microscopy; this is 'superresolution', or resolution better than predicted by a simple form of Abbe theory. Under some conditions, for example when using relatively large pin-holes, the resolution in confocal scanning may be strictly speaking no better than in conventional microscopy, but seems to be superior because of the relatively high contrast, not obscured by hazy, out-of-focus detail.

Resolution depends in conventional fluorescence microscopy entirely on the aberrations and numerical aperture of the imaging optics, and the wavelength of the emitted light. In fluorescent confocal scanning, however, resolution is affected by the wavelength and the optics (including pin-hole size) on both the illuminating and imaging sides. The best results are obtained when the exciting light has the same wavelength as the emitted light, but in practice the exciting radiation is usually of shorter wavelength than the emitted light except in the special case of two-photon fluorescence (see p. 79). The resolution in confocal fluorescence scanning therefore tends to be determined mainly by the optics and wavelength on the exciting side of the system.

With a grating object, fluorescence confocal scanning gives on the Sparrow criterion a lateral resolution up to twice as good as in conventional microscopy (McCutcheon, 1967; Cox et al., 1982). This is confirmed by the computer simulation (Goldstein, 1992). This indicates K values of 0.25 with fluorescence confocal scanning and 0.5 with conventional fluorescence microscopy on the Sparrow criterion, and values respectively of about 0.38 and 0.53 on the Rayleigh criterion. In resolving a pair of isolated test objects fluorescence confocal scanning has a less marked advantage over conventional fluorescence microscopy, the two methods having K values of about 0.32 and 0.44 respectively on the Sparrow criterion, and 0.41 and 0.48 on the Rayleigh criterion.

The lateral resolving power in transmitted-light confocal scanning microscopy is only marginally better than in conventional transmitted-light microscopy. Using grating objects, conventional transmitted-light full-cone (effectively incoherent) illumination and confocal scanning both have K values of 0.5 on the Sparrow criterion (Goldstein, 1992). Confocal scanning does however give somewhat better contrast, so that on the Rayleigh criterion the corresponding K values are respectively 0.53 and about 0.515. For comparison, transmitted-light coherent illumination at an obliquity just matching the objective aperture has K values for grating objects of 0.5 on both the Sparrow and Rayleigh criteria.

Using the Rayleigh criterion for resolution, according to Wilson (1990a p. 29) the distance between two closely spaced points which are just resolved by transmitted–light confocal scanning is 8% closer than is possible with conventional microscopy, and 32% closer than with a conventional microscope using fully coherent illumination. This is confirmed by the computer simulation (Goldstein, 1992); confocal scanning and conventional full-cone illumination both have a K value of 0.44 on the Sparrow criterion, and have K values of 0.48 and 0.5 respectively on the Rayleigh criterion. Conventional microscopy with a condenser aperture somewhat larger than that of the objective incidentally has the same K value (0.48) as confocal scanning on the Rayleigh criterion, and rather better resolution (0.375) on the Sparrow criterion.

12.3.3 Depth resolution. Excellent depth discrimination is found in fluorescent and even more so in transmitted-light confocal scanning systems (Wilson, 1990a), and is probably the main reason for the popularity of confocal scanning in cell biology.

In conventional microscopy out-of-focus detail appears blurred in the image, but the 'sectioning' property of confocal scanning is such that parts of the object slightly out of focus tend to be rejected altogether (Wilson, 1990a, p. 14). Extremely sharp, high-resolution 'optical sections' can therefore be obtained through relatively thick specimens, practically unaffected by blur from object details lying above or below the plane of best focus (White *et al.*, 1987). The qualitative explanation for this is quite simple: an object point not in the plane of best focus is illuminated relatively weakly (Figure 12.2), and since light from the point is distributed fairly widely over the image plane, very little energy passes through the pin-hole to the detector.

Wilson (1990a, p. 11) measured the change of focus necessary to reduce the intensity of the transmitted-light confocal image of a plane object by one half. He found the 'strength of sectioning' to increase not quite linearly with the objective NA. For objectives of a given aperture the sectioning property was somewhat more marked with a dry than with an oil immersion lens (see also Majlof and Forsgren, 1993). Plane, point and line objects behave somewhat differently, for example point objects show a weaker sectioning property than plane ones.

An alternative theoretical approach to the problem of depth of focus and/or strength of sectioning (Kimura and Munakata, 1990), unfortunately not readily applicable experimentally, is to find the smallest vertical distance apart at which two overlapping objects can still be distinguished as separate by careful focussing. Assuming the Sparrow criterion (i.e. unlimited sensitivity of discrimination) and incoherent (e.g. fluorescence) imaging, this is equivalent to finding the change of focus necessary to reduce the central intensity of the image of an object to one half of that found at the plane of best focus.

Two fairly simple methods are commonly used to combine confocal scans of a three-dimensional object, made at different (z-axis) planes, to form images with an extended apparent depth of focus. In the 'extended focus' method the successive images are added to give a composite image, while in the 'auto focus' method each point of the final image is given the maximum intensity recorded for the point during the axial scan. It is said (Sheppard and Cogswell, 1990) that auto-focus images usually exhibit better contrast than extended focus ones, but that the latter are superior in filtering out noise.

12.4 NON-IDEAL CONFOCAL SCANNING

12.4.1 The effect of lens aberrations. In confocal scanning, axial imaging properties are in general more sensitive than lateral imaging to lens aberrations (Sheppard and Cogswell, 1990). High-resolution confocal systems tend to be especially sensitive to spherical aberration, which can be due simply to a poor lens but can also be caused by a faulty microscope tube length or the use of a coverglass of incorrect thickness. Focussing deeply into a specimen can introduce spherical aberration, for example when using an oil-immersion objective and a specimen mounted in an aqueous medium. Chromatic aberration can also be troublesome, particularly in fluorescent confocal microscopes using different wavelengths for the exciting and emitted radiation (Wilson, 1990b).

12.4.2 The effect of scanning-spot size. The resolving power of confocal microscopy is naturally impaired if the exciting and detecting apertures are too large, i.e. the system is not diffraction limited, and with very large holes approaches that of conventional microscopy. Depth discrimination is considerably less sensitive than lateral resolution to pin-hole size (Wilson, 1990b), and good images can be obtained (Figure 12.3) using pin-holes of moderate size (Wilson and Carlini, 1988; Shuman, 1988). Especially in fluorescence systems it is often advantageous to improve the signal/noise ratio by using reasonably large spots, and the use of a slit-shaped detector instead of a symmetrical one may be acceptable in view of the improvement in the signal despite the accompanying asymmetrical deterioration in resolving power. It is sometimes possible to improve both the lateral and the axial resolution of a confocal system by scanning consecutively with detectors of different sizes, and subtracting one result from the other – this effectively allows extrapolation of the results to compute the results which would have been obtained with a smaller spot size.

The Zernike program can be used to study the depth of focus in confocal fluorescence microscopy according to the definition used by Kimura and Munakata (see above). Assuming a minimum-diameter absorbing object, and identical objective and condenser spot sizes and numerical apertures, under some conditions of defocus or spot size images showed a central dip in intensity. With a given numerical aperture the depth discrimination is worse with larger measuring spots, and this effect is more marked with lenses of high aperture. For a given width of measuring spot, the depth discrimination is mostly worse with objectives of lower aperture, but there are some anomalies – for example, with a finite spot size depth discrimination appears to be somewhat better with a numerical aperture of 0.75 than with the (purely theoretical) maximum NA of 1.0.

12.5 COMBINATION OF CONFOCAL SCANNING WITH OTHER MICROSCOPIC TECHNIQUES

Confocal scanning microscopes currently available commercially mostly have incident-light reflectance or fluorescence optics. Reflectance instruments are typically used in materials science and the semiconductor industry, while in biology most confocal scanning microscopes are fluorescence models used together with a wide variety of fluorescent dyes (fluorochromes). Such instruments generally have a relatively shortwave laser as light source and perhaps 'crossed' exciting and barrier filters in suitable positions in the optical path.

Confocal scanning microscopes can however be combined with almost any microscopical technique including the use of polarized light, apodization, an annulus in the back focal plane of one or other of the imaging lenses to improve depth discrimination (Sheppard and Cogswell, 1990; Sarafis, 1990), interference methods of various types and fluorescence ratio-imaging. The advantages of confocal scanning and of two-photon fluorescence (p. 79) are often complementary. When using digital imaging techniques quantitation of the results is relatively easy, as is image enhancement by such methods as subtracting an image obtained under one set of conditions from that obtained under another.

The huge variety of applications will doubtless increase further as versatile and affordable apparatus increasingly becomes available.

12.6   EXPLAINING 'SUPERRESOLUTION' IN CONFOCAL SCANNING

The resolution obtainable with an optical instrument was studied by Toraldo di Francia (1955) in terms of the number of degrees of freedom in the object and in the image. An instrument capable of sampling data points at a certain frequency produces an image with a finite number of degrees of freedom, which can correspond to an infinite number of different objects. The situation can however be saved by a priori knowledge – for example, knowing that the object consists of either one or two points amounts to having '... an infinite amount of information about the object and needing only one additional bit of information ...'. For '... a super-resolving pupil to be useful, the observer must know beforehand that the whole field is black outside a small region about the centre ...'.

Cox and Sheppard (1986) somewhat similarly consider the information-carrying capacity of any optical system to be invariant, and optical information to have three spatial dimensions, one temporal dimension, and two independent states of polarization; resolution can be increased at the expense of other factors by increasing the time taken by the scanning system to form the image, provided it is known that the object does not change with time.

Many practical microscopists may find the above approach rather abstract, and prefer a more direct approach to the problem. A simple, qualitative explanation is given above (p. 87) of the reduction of blurred out-of-focus detail in confocal scanning, but a more rigorous explanation of the advantages of confocal scanning is probably impossible without an extensive use of mathematics. Perhaps we have to be satisfied with understanding that the presence of two pin-holes in confocal scanning somehow alters the diffraction-limited image of an object point, as has been shown theoretically and experimentally by various workers. For example, according to Wilson (1990a, p. 26), the image of a point object has an appreciably higher proportion of the energy in the central maximum in confocal scanning than in conventional microscopy. In the confocal case using two equal lenses, '... although the functions have the same zeroes, the central peak of the confocal response is sharpened by a factor of 1.4 relative to the conventional image (measured at half-peak intensity. The sidelobes are also dramatically reduced ...'.

The Zernike computer simulation (Figure 12.3) similarly shows that the image of a fluorescent point object formed by conventional micro-scopy is substantially wider than with scanning confocal microscopy using ideally small detectors; scanning apertures 2 or even 4 wave-lengths wide also gave quite good results.

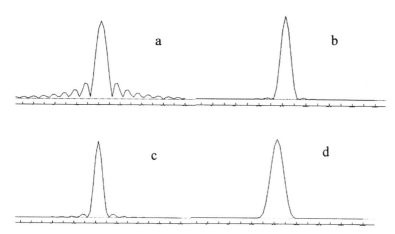

Figure 12.3 Point object imaged (NA 0.25) by conventional and confocal scanning fluorescence microscopy. Amplitude plots of (a) conventional fluorescence. (b) confocal (spot width 0.5 λ). (c) confocal (spot width 2 λ). (d) confocal (spot width 4 λ). All plots are normalized to unity maximum. Note the narrowing of the peak and the suppression of the side-lobes by confocal scanning, and the fact that an ideally small spot is not essential to obtain significant improvement.

12.7 CONFOCAL SCANNING SIMULATION IN THE ZERNIKE PROGRAM

Both transmitted-light and fluorescence confocal scanning can be simulated; the former assumes coherent and the latter incoherent imaging (see Section 3.8). The numerical aperture, focus, spherical aberration and glare of the objective are set by the user, together with the width (in half-wavelength units) of the measuring pin-hole. The corresponding characteristics of the illuminating optics are by default identical, but can be altered if desired. In fluorescence confocal scanning the ratio of wavelengths of the exciting and emitted light rays can be unity (the ideal condition), but the wavelength of the emitted light can be up to twice that of the exciting light.

Chapter 13

Microdensitometry

13.1 THE BOUGUER–BEER LAW AND A NOTE ON TERMINOLOGY

Densitometry (Hiskey, 1955) is the measurement of the amount of light-absorbing material in a beam of visible, ultra-violet or infra-red light, and microdensitometry is the application of densitometry to microscopic specimens. Both depend to a large extent on the Bouguer–Beer (Beer–Lambert) law.

If (Figure 13.1) a parallel beam of monochromatic light passes through a homogeneous light-absorbing object ('parallel', 'monochromatic' and 'homogeneous' are discussed later), some light is converted into heat or

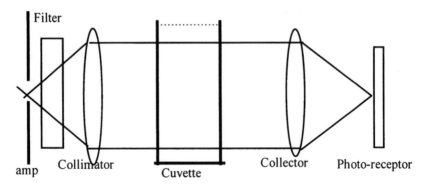

Figure 13.1   Simple densitometer

other forms of energy. The transmittance of the specimen is the dimensionless ratio

$$T = \frac{I_t}{I_0} \tag{13.1}$$

where I_0 and I_t are respectively the intensities of the beam on entering and leaving the object. The absorbance D is by definition

$$D = \log_{10}\left(\frac{1}{T}\right) = -\log_{10}(T) \tag{13.2}$$

whence

$$T = 10^{-D}. \tag{13.3}$$

According to the Bouguer–Beer law the absorbance of a solution of a light-absorbing solute is directly proportional to both the pathlength through the solution (this is Lambert's[1] or Bouguer's[2] law) and to the concentration of solute (this is Beer's[3] law). In mathematical form

$$D = -\log_{10}(T) = a.b.c$$

where b is the pathlength through the solution, c is the concentration of solute and a (the *absorptivity*) is a constant characteristic for the solute. If b and c are respectively expressed in cm and $gm.cm^{-3}$, since abc is dimensionless absorptivity has the dimensions cm^2gm^{-1}, or area per mass. The numerical value of the absorptivity depends on the identity of the solute, the wavelength of the light and the units in which b and c are expressed.

Why is there a logarithmic function instead of a simple ratio in the Bouguer–Beer law? Suppose that absorption in a solution reduces the intensity of a light beam by 50%. If a second, identical cuvette is placed in the optical path the intensity of the light will again be halved, to 25% of the original value, and with a third cuvette in place the transmittance becomes 12.5%. The number of cuvettes (0, 1, 2 or 3) is not linearly related to the corresponding transmittances 1.0, 0.5, 0.25 and 0.125, but is directly proportional to the absorbances 0.0, 0.301, 0.602 and 0.903 – hence the usefulness of the logarithmic function. Now imagine that using a single cuvette the concentration of a coloured solute is doubled. Provided the state of the solute is unaltered by the increasing concentration, twice as many solute particles are now present in the optical path, which is equivalent to having twice the thickness of the original solution. Again a logarithmic relationship applies.

[1] Johann Heinrich LAMBERT 1728–1777. Physicist, astronomer and philosopher of Alsace. Published his law in 1760, unaware of the previous work of Bouguer.
[2] Pierre BOUGUER 1698–1758. French hydrographer and physicist. Published his law in 1729.
[3] August BEER 1825–1863. Mathematician of Bonn.

13.1.1   A note on terminology.   Absorbance, absorptivity and transmittance have already been defined. *Optical density* is a somewhat oldfashioned synonym for absorbance. *Absorption* means $I_0 - I_t$, and is not the same as transmittance (I_t/I_0).

Densitometry implies estimating the amount of material present in a specimen by measuring its absorbance at a particular wavelength. In *spectrophotometry* (spectroscopy) the absorbance, transmittance or absorption spectrum of a specimen is constructed by making measurements at a series of different wavelengths in the hope of obtaining information on the chemical nature or state of the object.

13.2   ERRORS IN DENSITOMETRY AND MICRODENSITOMETRY

It is important to distinguish between *precision* and *accuracy*. Precision (reproducibility) is diminished by random variations and can be assessed by calculating the standard deviation of repeated measurements of a single object. In general precision can be improved to any desired extent simply by making more measurements on each object in a series. Accuracy, on the other hand, refers to the *correctness* of a measurement and can be affected by very subtle and elusive systematic errors. Accuracy is particularly difficult to assess and eliminate in microdensitometry, since the true value of what is measured is often not known in the form required, e.g. mass or activity of material per cell or unit projected area. Often the best that can be done in microdensitometry is to attempt to correct sources of error which are known to exist.

Many errors are common to both macro- and micro-densitometry, but some are peculiar to the latter. It should be emphasized that problems and potential errors found with mechanical, scanning microdensitometers (now sometimes regarded as being rather old-fashioned) in general apply also to systems comprising a microscope linked to a computer via a video camera.

13.2.1   The 'parallel beam' problem.   If the beam of light through a specimen is not parallel, the pathlength of different rays vary and Bouguer's law is inapplicable. There is no great difficulty in obtaining an effectively parallel beam of light in ordinary densitometry, but in microscopy (using either Critical or Köhler illumination) light passes through the microscopic object not as a parallel beam but as a cone with a finite numerical aperture. Some authorities accordingly recommend that in microdensitometry the NA of the illuminating light should be restricted to (say) 0.2 or 0.4, but detailed calculations (e.g. Uber, 1939; Blout *et al.*, 1950) indicate that such a precaution is seldom necessary.

There is no error with spherical objects, and the error is usually negligible even with relatively unfavourable specimens such as flat, extended sheets of coloured material.

13.2.2 Monochromator errors. According to physical theory a beam of light cannot consist only of a single wavelength. The bandwidth of a real beam is commonly defined as the range of wavelengths having an intensity at least half that of the maximum. Different wavelengths of light are in general differently absorbed by a coloured substance, i.e. have different absorptivities, so that if light passes through a coloured substance the more highly absorbed wavelengths disappear first and the proportion of other wavelengths increases. As a result the Bouguer–Beer law is not perfectly obeyed, and with a wide bandwidth the apparent absorbance may tend to a plateau value.

The most highly absorbed wavelength (λ_{max}) is usually used for the densitometry of relatively weak solutions and pale microscopic specimens, but with strongly absorbing material the transmittance at λ_{max} may be too low for accurate measurement. Densitometry requires light with a bandwidth narrow enough for the Bouguer–Beer law to be closely followed over a reasonable range of absorbances. This bandwidth depends on the absorption spectrum of the object, a narrower bandwidth being needed for measurements at or near a sharp peak of the absorption spectrum. Most dyes used in biology have spectra with fairly broad absorption peaks, so that their densitometry needs light with only a moderate bandwidth, say 10–15 nm.

Light for most microdensitometric purposes can readily be isolated from the almost continuous spectrum produced by an incandescent lamp. A good-quality continuous interference filter conveniently provides a suitable bandwidth at a wide range of wavelengths, although the bandwidth of some commercial filters is rather too large. Individual narrow-band interference filters give adequately monochromatic light but are inconvenient if a wide range of wavelengths is required. With all interference filters care must be taken to exclude invisible UV or IR light, which must sometimes be eliminated by an additional glass filter. Prism or grating monochromators have the advantage that the purity of their output can be controlled by the use of variable input and/or output slits. Bright quasi-monochromatic light can be isolated from the output of an arc lamp (e.g. a high-pressure mercury arc), in which much of the energy is concentrated into a few specific, narrow bands. One or more very narrow bands are usually present in laser light, but lasers are not very suitable for densitometry as the intensity of their output tends to be relatively unstable.

Bouguer's law can be tested in microdensitometry (Goldstein, 1975) by using a variable-path-length cuvette filled with a solution of the dye used to stain the specimen, inserted anywhere in the optical path of the microdensitometer (not necessarily in the normal position of the specimen). Provided the apparent absorbance is proportional to the pathlength over the range of absorbances likely to be encountered in the microscopic specimen, it usually does not matter if the proportionality factor is less than the ideal factor obtainable using quasi-monochromatic light. For example, haemoglobin has a narrow absorption band (the Soret band) at ca. 415 nm, but microdensitometry of red blood cells at this wavelength is satisfactory even with a relatively wide bandwidth (James and Goldstein, 1974).

Serious monochromator error can occur in microdensitometry using three-colour video cameras without additional colour filtering. Video cameras can respond to a huge number of different shades of light, but each of the three camera channels responds to too wide a range of wavelengths for Bouguer's law to be obeyed over a useful range of absorbances. The same *caveat* applies to the use of colour film in photographic microdensitometry (see below).

Adherence to Beer's law in microdensitometry is not readily investigated, as the whole concept of concentration is dubious in the case of dye molecules adsorbed on to an insoluble substrate (see below).

13.2.3 'Homogeneous' – randomness and distribution error. In practice, the light-absorbing components of many substances are not oriented randomly and as a result do not behave in the ideal manner specified in Beer's law. For example, the absorbance of alcoholic solutions of the dye Methylene Blue is proportional to the concentration over a wide range, but the dye tends to be dispersed in dilute aqueous solutions and aggregated in concentrated ones. As a result the colour and absorptivity of aqueous solutions of Methylene Blue change with concentration, and Beer's law is not valid. Such behaviour is often referred to as 'a deviation from Beer's law', but it is better to say that in this instance the object is not sufficiently homogeneous for the law to apply.

A special case is of some importance. If a substance exists (say) in a monomolecular form in dilute solutions but aggregates into dimers with a different absorptivity as the concentration increases, at one particular wavelength (the *isosbestic point*) the absorptivities of the monomer and dimer will be identical. Densitometry at this wavelength yields information about the total amount of dye present irrespective of the extent of aggregation.

Distribution error is a type of non-homogeneity crucial in microdensitometry. Consider a conventional, transmitted-light microscope (Figure 1.2). By simply adding a colour filter and a photoreceptor a simple instrument can be improvised with which it is possible to measure the transmittance and hence the absorbance of a uniform, extended object.

This is the whole-field or 'plug' method of microdensitometry. Consider, however, that an illuminated microscopic field is uniformly filled with coloured material having a transmittance of (say) 0.1 and a corresponding absorbance of $-\log_{10}(0.1) = 1.0$. If the same amount of material is condensed into half the area, the concentration and hence the absorbance are doubled so that the dense part of the field has an absorbance of 2.0 and a transmittance of 10^{-2}. Since the empty half of the field has a transmittance of 1.0, the whole field has a mean transmittance of 1.01/2 and an absorbance of $-\log_{10}(0.505) = 0.3$ instead of the original value of 1.0. This *distribution error* is due to the non-random character of the specimen.

Distribution error in microdensitometry can be tackled in various ways. In *photographic microdensitometry* chemical or spectrophotometric measurements are made of silver (Niemi, 1958) or dye (den Tonkelaar and van Duijn, 1964) eluted from a piece cut out of a photomicrograph. *Two-wavelength microdensitometry* (Garcia and Iorio, 1966; Mendelsohn, 1966) depends on whole-field transmittance measurements made at two different wavelengths, while in the *two-area* method (Garcia, 1965) whole-field measurements are made using a single wavelength but two different areas of illuminated field; from these measurements it is possible to calculate integrated absorbances (partly) corrected for distribution error. The most elegant and generally applicable method to reduce or eliminate distribution error is however *scanning and integrating microdensitometry*, and the further discussion deals almost exclusively with this method.

In scanning and integrating microdensitometry the transmittance is measured of an area small enough to be effectively uniform, so that the corresponding computed absorbance is effectively free from distribution error. The process is repeated as the measuring spot is moved (*scanned*) in a systematic pattern (*raster*) over the object. All the individual absorbances are added (*integrated*) and finally displayed as the *integrated absorbance*, which equals the projected area of the object multiplied by its mean absorbance. To ensure that the whole discrete object is measured the scanned area should include a rim of empty background. If this has a transmittance of unity it will not contribute to the integrated absorbance. Note that integrated absorbance is not the same as the product of the area and an absorbance calculated from the average transmittance, as would be the case with simple whole-field microdensitometry.

The measuring spot can be the reduced image in the specimen plane of an illuminated aperture situated in a conjugate plane. Relative movement of the spot and specimen is then achieved by moving the object on a motorized stage, or by moving the 'flying' spot by means of vibrating mirrors or prisms between the specimen and the physical aperture. Alternatively the measuring spot can be a moving circular or rectangular aperture, or successive elements of a video camera, situated in an optical plane conjugate with the specimen. Moving the diffraction-limited image of a spot over a real object is optically equivalent to moving a physical aperture over the diffraction-limited image of a specimen (see an analogous discussion on confocal microscopy).

Ideally the measuring spot should be infinitely small. Due to diffraction it is of course impossible for the effective size of the measuring spot to be smaller than an Airy disc, and in practice the spot is often rather larger than this in order to allow through more light to the photoreceptor and thus improve the signal/noise ratio of the system. With a measuring spot of finite width the apparent integrated absorbance of an isolated or non-uniform object is reduced by residual distribution error, and a similar effect is produced if the size of the measuring spot, referred to the object plane, is enlarged because the object is out of focus. Methods have been described for the microdensitometry of biological sections thicker than the depth of focus of the objective, but for the most critical work the whole specimen should be accurately in focus, so that flattened cells in smear preparations are ideal.

According to elementary theory based on geometrical optics (i.e. ignoring diffraction), the distribution error of a one-dimensional line scan over the edge of an extended object is directly proportional to the width of the measuring spot (Goldstein, 1971). In the case of a two-dimensional scan which completely covers a discrete round object, as the measuring spot size is increased greatly or as the object is thrown further out of focus the apparent integrated absorbance tends not to zero but to a limiting value of $0.4343A(1 - I_t)$, where A and I_t are respectively the area and the true transmittance of the specimen.

Experimentally it can be shown (Goldstein, 1971) that with measuring spots of small or moderate size the relative error in the integrated absorbance is greater with objects which are small and/or of high absorbance. The error is approximately proportional to the width of the measuring spot, so that by making measurements with spots of two or more different sizes it is possible to estimate the true integrated absorbance corresponding to a measuring spot of zero size.

The predictions of a more sophisticated theory which takes diffraction into account (Goldstein, 1982b) are identical with those of geometrical optics for relatively large objects and measuring spots. Diffraction effects however become relevant with objects or measuring spots comparable in size with the wavelength of light, and there is theoretically little or nothing to be gained by using a measuring spot with a geometrical width less than about one-quarter of the wavelength of the light used. The error in the apparent integrated absorbance obtained by a one-dimensional scan over the edge of an extended object can be shown to be approximately proportional to the measuring spot width provided this is greater than about three times the wavelength of light. With very small spots, of a size comparable with the wavelength of light, there is always at least a little distribution error per line scan. According to diffraction theory there is often significant error in the apparent integrated absorbance obtained by a two-dimensional scan over a discrete object of relatively high absorbance, but there is negligible distribution error with very pale objects. Distribution error can therefore be controlled even with very small objects by using an off-peak wavelength, but the use of very low absorbances is limited in practice by noise (random variations) in the system.

In the M85 scanning microdensitometer formerly made by Vickers Instruments, a round aperture imaged into the object plane as the flying spot functioned also as the output slit of the prism monochromator. Changing the diameter of the aperture therefore simultaneously affected both the spatial resolution and the monochromator bandwidth. Unless the optics of a given instrument are thoroughly understood, unexpected features of this sort can vitiate the assessment and correction of systematic errors.

13.2.4 Glare (stray light). Scattering of light results in diffuse spreading of light over the image plane (i.e. *glare* or *flare*), and in a well-maintained microscope is mainly due to multiple reflections at glass–air interfaces in the objective lens. This is analogous to the so-called Schwarzchild–Villiger effect first described in astronomy. Glare reduces apparent absorbances and integrated absorbances, particularly in the case of relatively dense microscopic specimens (Naora, 1951). For example, in the presence of uncorrected glare a small lymphocyte nucleus darkly stained for DNA by the Feulgen method may have an apparent integrated absorbance which is significantly (perhaps 5–10%) less than that of a large, palely stained monocyte nucleus, even though the two cell types probably contain the same amount of both DNA and dye (Bedi and Goldstein, 1974, 1976).

Glare can be estimated by measuring the apparent transmittance in the centre of an opaque particle such as a carbon granule, big enough (several micrometres wide) so that diffraction effects in its centre can be ignored, and ideally about the same size as the discrete object to be measured. Glare figures of 5–10% are typically found with an unselected lens and a large illuminated microscopic field, but glare can be reduced by careful selection of objectives, scrupulous cleaning of accessible surfaces, and especially by restricting the illuminated area as much as possible. In stage-scanning microdensitometers the illuminated area need be only a little larger than the measuring spot so that glare can be almost eliminated (cf. confocal scanning). With other types of microdensitometer the area illuminated cannot be smaller than the object to be measured, but by closing the field iris as much as possible glare can usually be reduced to perhaps 2%. In critical work even this amount of glare should be corrected. If G is the glare measured under defined conditions, the true absorbance of a specimen of apparent transmittance T_{ap} measured under the same conditions is given by

$$D = \log_{10}\left(\frac{1 - G}{T_{ap} - G}\right) \tag{13.4}$$

It is therefore possible to compensate for glare by subtracting a small current or voltage proportional to G from the transmittance signal of both the empty field and the specimen. This is conveniently done automatically in a way analogous to the routine offsetting of the dark-current of the photomultiplier tube (Goldstein, 1970b).

Glare in a microscopic image is due solely to scattering of light between the image plane and the object, but scattering of light between the object and the lamp, especially in the microscope condenser, contributes to glare in the final image of the field stop. In one commercial microdensitometer glare was corrected automatically by offsetting the transmittance signals to make the final image of the field stop appear black, but since only objective glare is relevant in microdensitometry this procedure can give serious overcorrection.

13.2.5   Linearity of photometric response.   To obtain the transmittance T it is necessary to measure I_0 and I_t accurately. This requires that the photoreceptor response in volts, amps or other units is linear over a fairly wide range of intensities, and although this is in general true of photomultiplier tubes it is not necessarily the case with other types of measuring device. In practice it is prudent in preliminary experiments to establish the linearity of a densitometer by plotting the response of the detector to different intensities of light.

Once the transmittance has been accurately measured the absorbance must be calculated. This can be done by an operational amplifier which automatically processes the output signal of the receptor, or by using an electronic calculator or computer.

In establishing how much light has been absorbed by a solution it is necessary to know the intensity of the beam of light both before and after it passes through the object. It is not sufficient simply to measure the intensity of the beam with and without the specimen-containing cuvette in place, since some energy is in general lost not only by absorption by the solute but also by absorption in the glass of the cuvette or the solvent itself, and especially by reflection at glass-air and glass-solution surfaces. It is usually necessary to estimate I_0 by measuring the intensity of light which passes through a *blank*, for example an identical cuvette containing pure solvent. In microdensitometry it is customary to use empty background near the object of interest to adjust the 100% transmittance level – this may be done in various ways, e.g. by altering the voltage of the lamp supply, the use of neutral filters in the optical path, changing the exit slit of the monochromator or controlling the 'gain' of an electronic amplifier.

Photo-sensitive devices invariably produce a small output (*dark current*) even in the total absence of light, which can result in significant systematic error in densitometry. The dark current must therefore be off-set electronically, or compensated for by subtracting some value from every transmittance reading.

13.2.6 Noise. 'Noise' in a measuring system consists of random electrical impulses due to unavoidable imperfections in the electronics or even to quantum fluctuations when using very low light levels. Very low and very high absorbances are both unfavourable for precise measurements, and it can be shown (Hiskey, 1955) that with extended objects the best absorbance for repeatable densitometry is around 0.43.

Random variations can often be 'averaged out' by making repeated measurements, but under some circumstances lead to systematic errors (Goldstein, 1986). Thus densitometers are often designed to eliminate occasional very high and possibly aberrant absorbance readings, corresponding to very low transmittances. If (say) any transmittance less than 0.05 is registered as 0.05 precisely, in a transmittance signal with a true mean at or just above 0.05, part of the signal will due to random fluctuations be below the critical threshold. This results in an incorrectly high apparent mean transmittance and hence an incorrectly low apparent absorbance. Similarly, some instruments are unable (or unwilling!) to record negative absorbances, corresponding to apparent transmittances greater than the nominal value of 1.0. If only positive absorbances are recorded, in the presence of noise a true mean transmittance of 1.0 will be converted to an finite, positive absorbance instead of the true value of zero.

13.3 CALIBRATION, STANDARDS, NON-UNIFORM BACKGROUND

Microdensitometers in general display integrated absorbances in arbitrary machine units. These can and should be converted to absolute units with the dimensions of μm^2 by multiplication by a calibration factor, established experimentally using the same instrumental settings (magnification, wavelength, monochromator slit-width, scan rate etc.).

One of the best methods to obtain reproducible results is to use an internal standard. For example, to investigate the mean DNA content per nucleus of a given species, samples can be prepared, stained and measured together with those of a standard species with a well-established nuclear DNA value. Expressing the experimental results as fractions or multiples of the standard simultaneously controls both staining and instrumental variations.

Ideally an empty microscopic field would have a perfectly uniform intensity. In practice, imperfect adjustment together with inherent defects and dirt in the optical system invariably result in some spatial variation in brightness, which can affect the apparent integrated absorbance of specimens in different parts of the field. To overcome this it is good practice always to use only the centre of the microscopic field, and to subtract the apparent integrated absorbance of an empty area from that of the same area containing the specimen of interest.

13.4   APPLICATIONS OF MICRODENSITOMETRY

It is possible in principle to measure inherently coloured or stained material in almost any microscopic object including biological tissues, cells and intracellular organelles. An example of a naturally coloured material is haemoglobin, which absorbs both green light (hence the red colour of blood) and short-wave blue light. Nucleic acid strongly absorbs UV light at about 266 nm, and although it is normally colourless in visible light, DNA can be specifically stained red by the Feulgen[1] method. The quantity of DNA in nuclei, chromosomes and even parts of chromosomes can therefore be estimated indirectly from microdensitometric measurements of the attached dye. Other important biological applications of microdensitometry include the measurement of (inherently coloured) pigment, silver grains in autoradiographs, and stained enzymes, antigens and antibodies.

13.4.1   Errors associated with the specimen.   The Bouguer–Beer law assumes that loss of light in the specimen is due solely to absorption, i.e. conversion of light into another form of energy. If however the refractive index of the specimen is not the same as that of the surrounding medium, scattering of light can reduce the apparent transmittance. Adequate matching of the refractive indices of specimen and mounting medium is not difficult when using visible light, but scattering can be a serious problem with UV light. This is partly because the intensity of scattering is inversely proportional to the fourth power of the wavelength, so that scattering at (say) 266 nm, the wavelength of maximum intrinsic absorption of nucleic acid, is almost twenty times more serious than when using green light of ca. 550 nm. Furthermore, it is difficult or impossible to find mounting media which at 266 nm are transparent and

[1] Robert Joachim Feulgen 1884–1955, German biochemist and histochemist.

have the same refractive index as nucleic acid. This factor, rather than lack of adequate instrumentation, is probably the main reason why the microdensitometry of unstained nucleic acid is no longer popular despite its early introduction by Caspersson and his school (Caspersson, 1936).

Exposure to bright light can result in fading of biological dyes, which is proportionally worse with pale, extended specimens than with dark, compact ones (Goldstein, 1981). Material intended for microdensitometry should therefore be stored in the dark and measured soon, and if possible a fixed time, after preparation.

Probably the most difficult problem in microdensitometry is to establish that a given staining method is both specific and stoichiometric. A staining method may be highly selective, i.e. demonstrate only a few components in a microscopic preparation, but to be called specific it is necessary (but not sufficient!) that its mechanism is thoroughly understood. Stoichiometry, or the proportionality of stain uptake to the amount of stained substrate present, is even more difficult to demonstrate rigorously, and very few of the innumerable staining methods to be found in the biological literature can be regarded as suitable for microdensitometry. It may be noted in passing that similar problems are found also in quantitative fluorescence microscopy, the difficulties of which have often been underestimated.

13.5 COMPUTER SIMULATION OF MICRODENSITOMETRY

Microdensitometry is accessed as an option in the main menu of the Zernike program, and can be combined with coherent, partially coherent or incoherent illumination. The parameters to be defined as (as usual) various properties of the objective, and in addition the width of the measuring spot and the positions of the left and right sides of the measured zone.

The Zernike program is essentially one-dimensional so that scanning over a two-dimensional discrete object cannot be simulated, but line scans can be made across the edge of an extended object, or over a round, prismatic or square discrete objects. The effect of the width of the measuring spot on distribution error can be investigated using objects of different transmittance. With the minimum width of measuring spot (a half-wavelength) some distribution error may still be present due to the image being affected by a finite objective aperture. The effects of glare, focus and spherical aberration on the apparent absorbance and integrated absorbance can also be demonstrated, as can the effect of the field stop in reducing error due to glare.

Chapter 14

Polarized Light (The Nicol Program)

14.1 GENERAL COMMENTS ON POLARIZED LIGHT

The following introductory account of polarized-light microscopy is very condensed and incomplete; for fuller discussions see e.g. Chamot and Mason (1958), Bennett (1950), Hallimond (1970), Hartshorne and Stuart (1970), Oster (1955), Shurcliff (1962) or Robinson and Bradbury (1992).

As stated previously, light waves are characterized by wavelength, speed, intensity and direction. Speed divided by wavelength equals frequency, while the amplitude equals the square root of the intensity. In addition to these properties light waves are *transverse*, unlike the longitudinal waves of sound. This means that the direction of the vibrations of light waves is at right angles to the direction of movement of the energy. If one could look at a light wave 'end-on' the vibration would appear to be at some ('azimuth') angle relative to a reference mark, for example in Figure 14.1 vibration OB is at an azimuth angle BOA relative to vibration OA.

Any individual wave-packet of light (or *photon*) is polarized, but in ordinary light, say from the sun, all possible azimuths of polarization are found and because there is no preferential azimuth the light as a whole is *unpolarized*. Reflection at an angle off certain dielectric (non-conducting) surfaces such as glass, water and tarmac can be somewhat selective for the plane of polarization, so that the proportion of the reflected light vibrating horizontally is altered relative to that vibrating vertically. Such

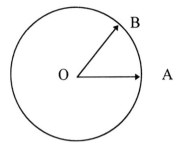

Figure 14.1 Light wave seen 'end-on'

reflected light is *partly polarized*. This happens at the highly curved edges of some components of high-powered lenses, and considerably affects polarized light microscopy (see p. 110). Light can also become polarized by selective *absorption* in a dichroic medium (see p. 112), by passage through a device such as a *Nicol prism* (Figure 14.2) which depends on total internal reflection of polarized light vibrating in a particular azimuth (the 'ordinary' ray) and the passage undeviated of the remainder (the 'extraordinary' ray), or by *scattering* by very fine particles. Thus blue light from the sky which has been scattered by molecules or fine particles in the upper atmosphere is partly polarized, and can be selectively blocked to some extent by dichroic spectacles.

Polarized light can be represented by one or more *vectors*, the direction and length of which respectively represent the azimuth and the amplitude of the vibration. A vector can always be resolved into two components (or vice versa*)* by the triangle or parallelogram of forces. For example, we can represent linearly polarized light of unit intensity emerging from a polarizer, by a vector of amplitude 1.0 at an azimuth angle of 0° relative to the transmission axis of the polarizer. On striking an object oriented at ±45° relative to the polarizer this light is resolved into two components, each (according to the theorem of Pythagoras) of amplitude $1/2^{1/2}$, at azimuth angles of +45° and −45° respectively. Note

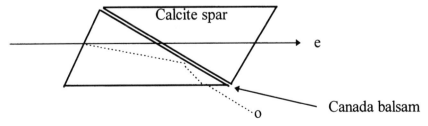

Figure 14.2 Nicol prism – elimination of 'ordinary' ray by total internal reflection

that intensity has been conserved, since the combined intensities of the components is $(1/2^{1/2})^2 + (1/2^{1/2})^2 = 1/2 + 1/2 = 1.0$.

If the object is a transparent, homogenous material such as unstressed glass, on emerging from the object the relative phase and amplitude of the two components are unaltered and they recombine into a single *resultant* identical in amplitude and direction with the light originally leaving the polarizer. If however the object is *birefringent*, that is to say the speed of light passing through it is a function of the direction of the light, the interaction of the components on leaving the object is a little more complicated. Birefringence is defined as the difference between the refractive indices parallel and perpendicular to a particular axis (the *principal* axis) of the material. Refractive indices and therefore also birefringence are dimensionless. Using quantitative polarized light microscopy we can measure the retardation (path-difference, phase-difference) between beams of polarized light. Retardation can be expressed as a fraction of a wavelength, as an angle (taking one wavelength as 360°), or as a linear measure in the unit used for the wavelength (usually μm or nm). For example, retardations of a quarter-wavelength ($\lambda/4$), 90° and 125 nm (i.e. 90/360 × 500 nm for light of 0.5 μm wavelength) are all equivalent. The retardation is proportional not only to the birefringence but also to the thickness of the specimen, and unless the thickness is known or can be measured the absolute value of the birefringence cannot be ascertained.

If two components at any azimuth happen to be exactly in phase with each other they can combine into a single, *linearly polarized* resultant. Polarized light beams of any relative phase can also interact, provided they can be resolved into components vibrating in the same azimuth. Calculation of the resulting vector or vectors starts off by working out the amplitudes of the components of the initial beams in the chosen azimuth, using the triangle of forces. The calculation must then take account of the relative phase of the interfering components, also using the triangle of forces but this time with the direction of the vectors representing not *azimuth* but *phase*.

In Figure 14.3 plane-polarized light of amplitude OA is incident onto a transparent object oriented at an azimuth angle of 45 degrees relative to OA. Vector OA can be resolved into components OB and OC, respectively parallel to and at right angles to the 'slow' axis OB of the object. Light vector OB is retarded by a phase angle Δ = FOD relative to OC. In turn, OB can be resolved into components OD and OE, respectively parallel to and at right angles to OA, and similarly OC can be resolved into components OF and OG. Being in the same azimuth components OD and OF can interfere with each other, as can OE and OG.

The resultant of OD and OF is vector OH, while that of OE and OG is OI. Note that OE and OG are 180° − Δ degrees out of phase with each

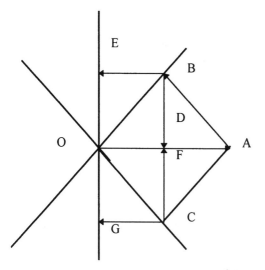

Figure 14.3 Resolution of plane polarized light into components (azimuth diagram)

other. If Δ had been zero, OE and OG would have been precisely half a wavelength out of phase since in their vectors face in opposite directions. If a birefringent object lies at an azimuth of ±45° relative to the polarizer, irrespective of the phase angle between the light components emerging from the slow and fast axes the final vectors OH and OI are always precisely a quarter of a wavelength out of phase with each other. The relative amplitudes of OH and OI *do* however depend on the value of Δ. In general, if there is some but not exactly the same amount of energy in both OH and OI the light is called *elliptically polarized*. If however Δ equals 180° *all* the energy is found in OI, while if Δ is 0°, exactly 360°, or some precise multiple of 360°, all the energy is found in OH. In these special cases the final resultant is linearly polarized light either at right angles to or parallel with the transmission plane of the polarizer. If at least some component of the linearly polarized light leaving the object is perpendicular to the plane of the polarizer some of the energy will pass through a crossed analyser and the object will appear bright on a dark background. Retardation of light by the object of a whole number of wavelengths will have no effect (but note that this is true only of monochromatic light, since with white light some colours will not be retarded by a whole number of wavelengths and will to a greater or lesser extent pass a crossed analyser.

Another special case is *circularly polarized* light. This consists of two components of *equal* amplitude, exactly 90° out of phase with each other. Such light has no preferential azimuth, since no matter how it is resolved

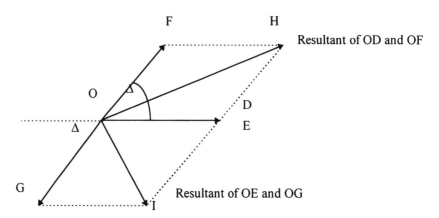

Figure 14.4   Recombination to give elliptically polarized light (phase diagram)

into components these will always be of equal amplitude, and 90° out of phase with each other.

14.2   POLARIZED LIGHT MICROSCOPE AND ACCESSORIES

A microscope equipped for polarized light work must have a polarizer below the specimen and an analyser above it, at least one of which should be rotatable. In modern microscopes these 'polars' are generally not traditional Nicol prisms, but are made of a dichroic material which allows through only light polarized in one particular direction (note that 'Polaroid', commonly used as a polar, is a trade-mark of the Land Corporation and not a generic term).

The specimen ought to be carried on a rotatable stage, and it should be possible to insert one or more compensators into the optical path. Many different types of compensator have been described (see e.g. Richartz and Hsü, 1949; Jerrard, 1948; Gahm, 1964) but not all are readily available commercially. In general, compensators are birefringent optical components which are used in such a way that their birefringence is equal in magnitude but opposite in sign to that of the specimen, so that the two cancel other out. The effective retardation of a compensator may be altered by pushing it along its long axis (in the case of a wedge), by rotating it around the optic axis of the microscope (e.g. a Köhler compensator), or by rotating it around its own long axis (e.g. a Berek compensator). A quarter-wave (Sénarmont) compensator acts in a slightly different way, and is in most applications kept stationary.

Some means of examining the back focal plane of the objective (*conoscopic examination*) is desirable. For quantitative work the rotating stage, rotating analyser and the compensator or compensators must be accurately calibrated. Objectives should be selected to be as strain-free as possible; for most purposes in polarized light microscopy ordinary achromatic objectives are preferable to fluorite or apochromatic lenses, which tend to suffer more from inherent birefringence due either to strain in their components or the use of inherently birefringent materials such as calcite.

14.2.1   The quartz wedge.   This is useful in the initial examination of an unknown object, to measure the retardation in whole wavelengths, or in fractions of a wavelength with relatively low precision. Fractions of a wavelength retardation are measured more accurately by other methods (see below).

It its simplest form the device consists of a wedge of quartz cut more or less parallel to its optic axis so that the retardation varies linearly along the length of the wedge. It is normally used at $\pm 45°$ between crossed polars in the slot of a Wright eyepiece, with the images of the wedge and object superimposed. If the retardation of the wedge exceeds one wavelength of monochromatic light a series of dark bars can be seen at right angles to the principal axis of the wedge, each dark bar corresponding to a place where the retardation is an integral number of wavelengths. If the retardation is engraved on the wedge itself the position of compensation can be read off directly with a precision of around a tenth of a wavelength.

If used with white rather than monochromatic light a quartz wedge at $\pm 45°$ between crossed polars shows a series of transverse coloured bands. Where the retardation is approximately 0.55 μm (the wavelength of green light), green light cannot pass the analyser and all the wavelengths which *do* pass to at least some extent combine to give a deep red (*red-of-first-order* or *deep Red I*). Towards the thin end of the wedge the colours vary (see Figure 14.5, based on data in Bennett, 1950) finally to become black (*black-of-first-order* or *black I*) where the retardation is negligible. From red-of-first-order towards the thick end of the wedge the (second order) colours visible include violet, blue, green and yellow. Orange-red is seen where the retardation is about 1.0 μm.

The upper part of Figure 14.5 shows intensity curves for three arbitrary wavelengths of light, plotted against distance along a wedge with a maximum retardation of 1000 nm (1 μm). The lower part of the figure shows the apparent colours of the wedge between crossed polars, using white light consisting of all visible colours with wavelengths between 0.4 and 0.7 μm.

If the wedge were *very* thick the apparent colour would be unsaturated or indistinct. Any wavelength is eliminated by the crossed analyser if when multiplied by any integer it exactly equals the retardation of the

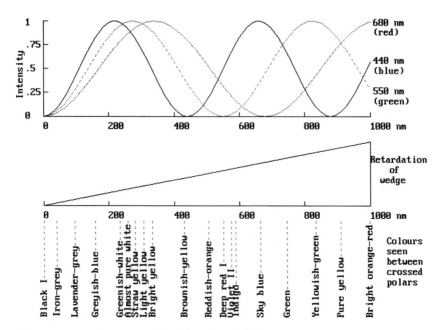

Figure 14.5   Quartz wedge with coloured light

wedge, but will pass the analyser if it equals the wedge retardation when multiplied by 1.5, 2.5, 3.5, etc. Intermediate wavelengths will pass to some extent. With a wedge retardation of (say) 5.0 μm, wavelengths of 0.625, 0.555, 0.5, 0.4545 and 0.4166 μm are completely eliminated and wavelengths of 0.666, 0.5882, 0.5263, 0.4762, 0.4348 and 0.4 μm are completely passed; the combined colour seen is a rather murky grey known as *'white-of-higher-order'*.

The *Soleil compensator* consists of two quartz wedges, one cut parallel to the optic axis of the quartz and the other cut perpendicularly, so that when superimposed they partially compensate each other and give a uniform field. By sliding one wedge longitudinally relative to the other, the combined retardation can be varied but remains uniform across the field. If suitably calibrated, a Soleil compensator can be used for quantitative or semi-quantitative work.

14.2.2 First-order red (sensitive tint) compensator. This is useful in qualitative work with white light to detect both the presence and the sign of birefringence.

Green light is retarded by one whole wavelength, blue light by somewhat more than a wavelength, and red light by somewhat less. Viewed against an empty field, at $\pm 45°$ between crossed polars, the compensator therefore gives extinction with green light but lets through some blue and some red light, resulting in a reddish or purple colour (see the discussion above on the quartz wedge). If the compensator is super-imposed on a moderately retarding specimen, which alone appears almost colourless, the combined colour is raised to blue if the slow axes of the compensator and object are parallel, and lowered to yellow if they are perpendicular to each other.

14.2.3 Elliptical (Brace–Köhler) compensator. This consists of a thin plate of birefringent calcite or mica, which retards the light by a small fraction of a wavelength (10° to 36°). It can be rotated around the optic axis of the microscope, and its effective compensation is a function of its azimuth angle (see p. 116 for a fuller discussion).

14.2.4 Quarter-wave (Sénarmont) compensator. Inserted parallel or orthogonal to the transmission axis of the polarizer, between the object (itself at $\pm 45°$) and the analyser, a quarter-wave ($\lambda/4$) compensator converts elliptically polarized light from a birefringent object into linearly polarized light, the azimuth of which is linearly related to the retardation of the object. This azimuth can be ascertained by rotating the (calibrated) analyser until the object appears maximally dark. Consider the elliptically polarized light emerging from a birefringent specimen oriented at $\pm 45°$ between crossed polars (Figure 14.4). Inserting a quarter-wave plate parallel to the plane of the polarizer eliminates the quarter-wave phase difference between components OH and OI (respec-tively parallel and orthogonal to the transmission plane of the polarizer). Components OH and OI can now interfere to give linearly polarized light at some azimuth angle HOR relative to the plane of the polarizer.

Some light passing through the birefringent object is therefore able to pass through the crossed polarizer, which must be rotated anticlockwise through the angle θ = HOR to restore extinction of the object. It can be shown that if θ is measured in degrees, $\theta/180$ equals the birefringence Δ of the object in wavelengths. The Sénarmont method is extremely useful for quantitative work up to one wavelength retardation. A significant drawback is its dependence on a compensator which is accurately quarter-wave for the light being used, although with a nominally

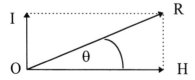

Figure 14.6 Rotation of azimuth of linearly polarized light (Sénarmont method)

quarter-wave compensator of $90 \pm 5°$ retardation, the measured retardation of a specimen has an error of less than about $0.1°$ (Goranson and Adams, 1933). It may be difficult to distinguish between a positively and a negatively birefringent object; since the same rotation of the analyser gives extinction of a positively birefringent object which retards the light by (say) $15°$, and a negatively birefringent one which advances the light by $360 - 15 = 345°$. Some prior knowledge, obtained for example by the use of a first-order-red plate or a quartz wedge, is therefore necessary when dealing with an new specimen.

14.2.5   Berek compensator.   This is a birefringent plate which can be rotated around its long axis, inserted at $\pm 45°$ between crossed polars. The retardation of the compensator is a function of the angle of rotation, and can be read off tables or a calibration scale on the device itself.

14.2.6   Wright eyepiece.   This consists of an eyepiece containing a rotatable analyser and a slot, in the front focal plane of the eyepiece below the analyser in a plane conjugate with that of the specimen, in which can be inserted a quartz wedge or a half-shade (e.g. Nakamura) plate (see below). The wedge compensator or half-shade plate is therefore seen superimposed on the image of the object, and the apparent darkness of different parts of the object can be seen simultaneously. In the computer simulation, a Wright eyepiece is automatically simulated if either the quartz wedge or the Nakamura plate is specified.

14.2.7   Nakamura half-shade plate.   The plane of polarization of linearly polarized light travelling more or less parallel to the optic axis of a piece of quartz is rotated through an angle proportional to the distance travelled. Quartz occurs naturally in two crystalline forms; 'right-handed' and 'left-handed' quartz respectively rotate the plane of polarized light clockwise and anticlockwise (if one looks in the direction in which the light is travelling).

The Nakamura plate consists of two thin pieces of quartz, one 'left-handed' and the other 'right-handed', cut in a plane at right angles to the optic axis of the quartz and mounted together with a sharp dividing line. The two types of quartz rotate the plane of polarized light passing through them in different directions, through a small angle which is usually between one and two degrees. The plate is normally used with a Wright eyepiece, so that the image of the plate is superimposed on the background or object. Using the plate is equivalent to rotating the analyser alternately clockwise and anticlockwise through a small angle, but the effects are seen simultaneously and not successively. If (say) the analyser is set so that the object is 'at extinction' (maximally dark) without the Nakamura plate, on inserting the plate the parts of the object on either side of the dividing line appear not black but grey, and of equal intensity. The human eye is not very good at judging or remembering *absolute* intensities, but *is* very good at *matching* intensities. By adjusting the analyser until adjacent parts of the object on opposite sides of the dividing line of the Nakamura plate appear equally dark, a very precise and reproducible setting of the angle of rotation of the analyser can be made. The Nakamura plate can also be used for setting the azimuth of a birefringent object, but is inferior to a Königsberger plate for this purpose (Hallimond, 1970).

14.2.8 Königsberger half-shade plate. This consists of two slightly retarding specimens (e.g. bits of cellophane or calcite), mounted together with a sharp dividing line and oriented with their 'slow' axes at a slight angle to each other. Using the plate is equivalent to inserting a slightly retarding compensator first parallel to and then orthogonal to the principal axis of an object, but with the advantage that the intensities on either side of the dividing line can be seen simultaneously. It is useful for detecting very weak birefringence and for setting the retardation of an adjustable compensator precisely, but is unfortunately not generally available commercially.

14.2.9 Bertrand lens. In order to examine the back focal plane of the objective in detail, one can remove the microscope eyepiece and insert a small telescope (often called a *phase telescope* because it is widely used in phase contrast microscopy) into the tube. The Bertrand (Amici–Bertrand) lens can be swung into place in the microscope tube below the existing eyepiece; together with the eyepiece it forms a telescope which can be focussed on the back focal plane of the objective. Such a device is useful not only in phase-contrast microscopy but also in polarized light work and even in ordinary bright-field microscopy, where it can be used when centring and adjusting the size of the condenser iris diaphragm. In the polarized light microscopy of crystals much information can be obtained by examining the pattern in the back focal plane of the objective, but this *conoscopic mode of observation* cannot be simulated by the computer program and will not be further discussed here.

14.2.10 Polarization rectifier. If polarized light is incident at an angle on a surface, the component polarized perpendicular to the plane of incidence is reflected more than is the parallel component, causing the plane of polarization of the (resultant) transmitted light to be rotated towards the plane of incidence. In a typical microscope objective this phenomenon affects light striking near the edges of curved lens elements, and markedly impairs the extinction between 'crossed' analyser and polarizer. In the back focal plane of the objective good extinction is seen only near the optic axis and in azimuths which are either more or less parallel with, or orthogonal to, the transmission plane of the polarizer. A 'Maltese cross' is seen, dark in the centre and in azimuths east-west and north-south, but relatively light in the diagonal axes. In the image plane the background extinction is poor, and contrast in the image is reduced.

This situation can be somewhat improved by antireflection coatings on all glass–air surfaces of the lens system, and by restricting the numerical aperture of the condenser so that most light is restricted to near the optic axis. The latter procedure however impairs the resolution. The effect was thought to be an unavoidable consequence of using high-powered objectives until Inoué and Hyde (1957) described the *polarization rectifier*. This consists of a thin, curved glass meniscus combined with a half-wave plate, ideally situated both between the polarizer and the condenser, and between the objective and the analyser. The meniscus introduces depolarization approximately equal in magnitude to that of the curved lens components, and the half-wave plate reverses the rotation so that depolarization at lens elements is effectively cancelled out and the extinction considerably improved. If only a single rectifier is used between the polarizer and condenser excellent background extinction is obtainable, but because beams diffracted by the object are incorrectly rectified a dark-field effect independent of birefringence may occur.

Although highly desirable for both qualitative and quantitative work with the polarizing microscope, the polarization rectifier is rather expensive to make and is not widely available.

14.3   TYPES OF ANISOTROPIC OBJECT

Anisotropic objects have properties which are different in different spatial directions. For the purposes of the present program, we can regard anisotropic objects as being birefringent, dichroic, or both birefringent and dichroic. It is assumed throughout that the object lies flat in a plane normal to the optic axis of the microscope, but in reality specimens are often tilted at an angle to the plane of the stage, which can make the polarized light analysis very complicated.

14.3.1   Birefringent objects. In ordinary isotropic (uniform) objects such as unstressed glass, the refractive index is identical irrespective of the direction through the object which the light takes. That is to say, the speed of light is the same in all directions in the medium. In anisotropic specimens, on the other hand, the speed of light and hence the refractive index depend on the direction of the light. This applies to such objects as glass under strain, crystals in which the chemical bonds are not symmetrically arranged in space, and many oriented biological macromolecules. Consider a fibre such as collagen, which consists of more or less parallel bundles of protein molecules. The speed of light is lower, i.e. the refractive index is higher, for light with its electrical component vibrating parallel to the long axis of the fibre. The fibre is *negatively birefringent*. In DNA the predominant orientation of the constituent nucleotides is more or less perpendicular to the long axis of the double helix; DNA fibres are *positively birefringent*, i.e. their refractive index is higher for light vibrating at right angles to the long axis of the whole molecule. Almost all oriented biological molecules are birefringent to a greater or lesser extent, and can therefore usefully be studied by polarized light microscopy. This is true for example of collagen, smooth and striated muscle fibrils, chromosomes, cilia, and actin in mitotic spindles.

The examples just considered are *uniaxial*, with only two different refractive indices. Some objects may have more than one optic axis, and light passing through them may have different refractive indices for perhaps three different directions. Such complicated cases are of importance in crystallography, but will not be further discussed here. Some important biological objects are however *radially symmetrical*. For example, starch granules and lipid droplets often have molecules arranged perpendicular to their surface; between crossed polars they typically give a *Maltese cross* appearance, with molecules at more or less 45° to the axes of the polars appearing bright, while those parallel or perpendicular to the polars are dark.

It is traditional to distinguish between *intrinsic* and *form birefringence*. Consider elongated, very small structures such as microscopic fibrils oriented parallel to each other and embedded in a medium of different refractive index. Such structures will appear birefringent between crossed polars, but the birefringence disappears if the mounting medium is replaced by one with the same refractive index as the fibrils. This is form birefringence. If however the fibrils consist of submicroscopic, elongated molecules with asymmetric electron shells, they will remain birefringent irrespective of the mounting medium – this is intrinsic birefringence. Frequently intrinsic and form birefringence occur simultaneously, and may even be of opposite sign. They may be distinguished by measuring the birefringence in a series of mounting media of different refractive index – intrinsic birefringence is what remains in the mounting medium giving minimum (or maximum) total birefringence.

14.3.2 Rotation of birefringence.

Some molecules, including quartz, some amino-acids and many types of sugar, can appear in either a left-handed or a right-handed form. If linearly polarized light enters an aqueous solution of a sugar such as dextrose, the plane of polarization is *rotated*. This phenomenon is used in a saccharimeter, but in biological sections the rotation of most substances is too small to be detected by ordinary methods (but see the Nakamura half-shade plate, p. 109)

14.3.3   Dichroic objects.

Consider a dichroic (but not birefringent) object oriented at 45° relative to the transmission plane of a polarizer (Figure 14.7). The plane-polarized light from the polarizer is resolved into components respectively parallel to and at right angles to the principal axis of the object. After transmission through the object, amplitude OB has been reduced by more than amplitude OA (the object has been assumed to be a linearly polarized resultant OR which is rotated through an angle relative to the original transmission axis of the polarizer. Part of this light could be resolved into a component *negatively* dichroic). Components OA and OB are still in phase with each other (remember that we assumed that the object was not birefringent), so can combine into at right angles to the polarizer, and would therefore pass through a 'crossed' analyser – the object would therefore appear bright between crossed polars.

The term *dichroism* applies to uniaxial objects, *pleochroism* to more complicated ones which can vary optically in several different directions. Dichroism can only be present in objects which absorb light, so that completely transparent specimens (although possibly birefringent) cannot be dichroic. Many transparent biological specimens can however be rendered dichroic by *staining*, by means of which elongated, dichroic stain molecules become attached to the anisotropic substrate in an oriented manner. A familiar example of this is the staining of the pathological substance amyloid by the dye Congo Red.

In contrast with birefringence, measurements of dichroism can in principle yield information about the degree of orientation of an object even if the absolute thickness of the specimen is unknown. Ordinary measurements using a simple polarizing microscope yield the ratio of the amplitudes of the light passing through an object parallel to and at right angles to its optic axis. This ratio *is* affected by the thickness, but if one measures the *absorbance* along the same two directions (absorbance is

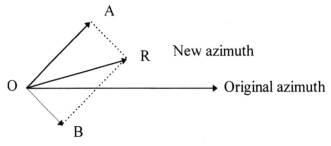

Figure 14.7   Rotation of azimuth of polarized light by a dichroic object

defined as minus the logarithm to the base 10 of the transmission), the ratio of the absorbances, i.e. the *dichroic ratio*, is a function of the orientation of the specimen but independent of its thickness.

If an object is dichroic (but not birefringent), and is oriented at $\pm 45°$ relative to the polarizer, the light coming from the object will be linearly polarized but have a changed azimuth angle. If, on the other hand (as is commonly the case), a dichroic object is also to some extent birefringent the emitted light will be elliptically polarized with the principal axes of the ellipse rotated relative to the original axis of the linearly polarized light coming from the polarizer. The analysis of such objects is naturally more complicated than that of purely birefringent or purely dichroic specimens but can be carried out if the proper apparatus is available (see below).

14.4 THE NICOL PROGRAM

The program quantitatively simulates polarized-light microscopy, and computes the apparent intensity of an anisotropic object and the empty medium in which it is mounted. It is assumed that the anisotropy is uniaxial, the object lies flat in a plane normal to the optical axis, the axis of maximum absorption (if the object is dichroic as well as birefringent) is parallel or orthogonal to the principal axis of the specimen, and the birefringence does not exceed ± 1 wavelength of the monochromatic light used. Diffraction and lens aberrations are ignored, and the object is assumed to be in focus.

The virtual polarizing microscope simulated in the computer program is more versatile than any commercially available instrument. It can be set up to have a polarizer, analyser, object and no fewer than three separate compensators, one below and two above the object. Generally elements 1 and 7 are respectively the polarizer and analyser, while element 3 is the object. Initially the compensators (components 2, 4 and 5) are respectively a quarter-wave plate, a quarter-wave plate, and a first-order red plate. The birefringence, dichroism and azimuth of all these components can be altered at any time. In addition, either a quartz wedge or a Nakamura plate can be inserted in a plane conjugate with that of the object. Only the azimuth and retardation of the quartz wedge can be changed; its transmittance is unity along both axes. Any or all properties of the object (azimuth, birefringence and dichroism) can be made random in order to practice the quantitative polarization analysis of an unknown specimen. A help function and a calculator are provided.

The program can simulate white light or monochromatic green illumination but not (at least at present) conoscopic microscopy or a Königsberger half-shade plate.

14.4.1   Actions of function keys in the program.   Some function keys (e.g. F1 and F10) operate throughout the program, while others only act in computer graphics mode (i.e. while a diagram is on the screen).

F1	HELP
F2	PEEK Examine current properties of polarizing components, and apparent intensities of object and background
F3	"In?" Insert or withdraw a component from the optical path
F4	rot.? Choose which component to rotate
F5	Rand. Randomise one or more object properties
F6	Menu< Jump to the main program menu
F7	Mouse Turn the mouse (if present) off or on
F8	Z/K Switch to the Zernike or Kohler programs
F9	.PCX Save graphics as a .PCX file
F10	Calc. Calculator
SHIFT/PRINTSCREEN	Dump screen to printer (DOS)
ALT/PRINTSCREEN	Dump screen to Clipboard (WINDOWS)
ESC	In main program: switch to the Zernike or Kohler programs or quit program altogether. While getting HELP: quit HELP.

14.4.2   The Calculator.   The Calculator is provided to help you assess unknown properties of a randomized object. With it you can evaluate (almost) any algebraic expression, such as LOG(3), SIN(PI) or 2*3. Permissible algebraic, trigonometric and Boolean functions include: ( ) AND OR EOR XOR = < = > = < > > < + − * / \ DIV MOD SQR ^ LOG EXP LN SIN COS TAN ATN RAD and DEG. Note that trigonometric operators expect radians not degrees – so first convert values in degrees into radians thus: SIN(RAD(45)).

Twenty-four case-insensitive variables from A to Z (excluding E) are available to store temporary results. For example, to store a calculated value in variable K type LET K = 2 * 3. Variables thus defined can be used in other parts of the program, e.g. to input data when changing values of polarizing components. In addition, 25 case-insensitive 'string' variables from A\$ to Z\$ can be used to store complicated expressions for later use. Thus LET B\$=−C*SIN(RAD(2*A)) will store the given expression in variable B\$. If later B\$ is typed, provided variables A and C have themselves been suitably defined the stored expression is evaluated and the result given.

Variables and functions are case-insensitive, and spaces and (often) brackets are optional. Thus a+sinpi and A+SIN(PI) give the same result, but the latter is clearer and is recommended. In some complicated expressions omitting the brackets gives an incorrect answer, as brackets can affect the order in which operations are carried out.

14.5 ANALYSIS OF AN UNKNOWN ANISOTROPIC OBJECT

14.5.1 Initial orienting of the crossed polars at exactly 90°. This is often necessary in practice, but not in the computer simulation (which assumes an ideal microscope!) unless you have already rotated either one. Remove all compensators and the object from the optical path. With the polarizer in position (usually east-west), rotate the analyser until the field is maximally dark. If a Nakamura half-shade plate is available, the setting can be made more precisely. The polarizer and analyser are now correctly crossed.

14.5.2 Orienting the object at ±45°. With crossed polars, but all compensators withdrawn from the optical path, rotate the stage until the object is maximally dark (use a Nakamura plate if available). Make a note of the stage setting, and rotate the stage by ±45°. The object is now correctly oriented. In the program, if the object azimuth has been randomized the absolute azimuth remains hidden. The azimuth of a rotating object relative to its original position is however shown. Rotating an object between crossed polars (and with no compensator in the optical path) gives a maximal brightness at ±45° absolute azimuth – make a note of how much the relative azimuth of the object differs from 45° when it is maximally bright, and add or subtract this value from the (unknown) azimuth of the object using Main Menu option 3. The object ought now to be correctly oriented; this can be checked by rotating the object again between crossed polars, using Main Menu option 5.

14.5.3 Testing for the presence and sign of birefringence. With the object at ±45° between crossed polars, insert a compensator of known retardation parallel to the object (i.e. at ±45° relative to the polarizer). For most biological objects a first-order red (sensitive tint) compensator is suitable if you are using white light illumination. If the interference colour is raised when the 'slow' axis of the compensator is parallel to the principal axis of the object, the object is probably positively birefringent. That is, the slow axis of the object is parallel to its principal axis. Conversely, if the interference colour is lowered, the object is probably negatively birefringent. If a Königsberger half-shade plate is available, this can also be used to assess the presence and/or sign of an object's birefringence.

14.5.4 Testing for the presence and sign of dichroism. The traditional test for dichroism is to withdraw all components from the optical path except for the object and either the polarizer or the analyser. Rotate (say) the polarizer. If the object appears relatively dark when the transmission axis of the polarizer is parallel to the principal axis of the object, the object is positively dichroic. Conversely, it is negatively dichroic if it appears darkest when the polarizer is perpendicular to the object's principal axis.

A more sensitive method (Goldstein, 1969b) is to rotate the analyser slightly (a few degrees only) first clockwise then anticlockwise from the crossed position, with the polarizer in place and the object at $\pm 45°$. In the presence of positive dichroism, the object will appear somewhat darker when the analyser is rotated towards the principal axis of the object and paler when the analyser is rotated in the opposite direction. Alternatively, with the object at 45° between crossed polars, insert a Nakamura plate into the Wright eyepiece. The part of a dichroic object under one side of the plate will look darker than the part on the other side of the dividing line – using the Nakamura plate on a known object will tell you how the plate behaves.

14.5.5 Using a quartz wedge compensator to measure whole wavelengths of retardation. This can be simulated by the program, using either monochromatic green or white light. With the object at $\pm 45°$ between crossed polars, insert a quartz wedge into the Wright eyepiece. A series of dark bands (some of which will be coloured if you are using white light) will be seen crossing the field. Where they cross the object the bands are deviated to one side or the other. The direction of deviation gives the sign of the birefringence, and the amount of deviation is a rough measure of the quantity. For example, if an identifiable band (say the black band using white light) is displaced by about one-and-a-half bandwidths over the object, the retardation is about 1.5 wavelengths. The fractional value can be more precisely measured by other means, e.g. a Sénarmont or Köhler compensator.

14.5.6 Analysis of birefringence with the elliptical (Brace–Köhler) compensator. The elliptical compensator is typically used for the precise measurement of small retardations (less than perhaps 0.1 wavelengths). The compensator consists of a thin birefringent plate with a retardation slightly greater than that of the object, which can be rotated about the optic axis of the microscope. The compensator is initially inserted with its fast axis parallel to that of the polarizer, and with the object at $+45°$ between crossed polarizer and analyser the compensator is rotated until the object appears maximally dark. If A is the angle through which the compensator has been rotated, the retardation of the object is O and the retardation of the compensator is C, then the retardation O of the object is given by the equation tan(0) = −2.tan(C/2).sin(2A). To work out the value using the Calculator, let C = the assigned retardation of the compensator (in degrees), let A = the measured angle (in degrees) through which the compensator has been rotated. Then the retardation of the object (in degrees) is given by DEG(ATN(−2*TAN(RAD(C/2)) * SIN(2*RAD(A)))). Note the use of deg and rad to convert from radians to degrees and vice versa – and be careful with the brackets! These equations are precise but rather complicated. A fairly good approximation, valid for small values of O and C, is −C*SIN(RAD(2*A)). Note that here both C and the final answer O are in degrees. With relatively highly retarding objects, the Brace–Köhler method can still be used. Insert the compensator at ±45° and rotate the object from parallel to the transmission axis of the polarizer until it is maximally extinguished or until it is at ±45°, i.e. parallel or orthogonal to the compensator. The greater the angle through which the object must be rotated, the lower is its retardation; this method only works if the retardation of the object is greater than that of the compensator. Let C be the retardation of the compensator, and A be the angle from parallel to the polarizer through which the object must be rotated to give extinction. The retardation of the object is now approximately given by C/SIN(RAD(2*A)). Compare this equation with that given above.

14.5.7 Analysis of birefringence with the quarter-wave (Sénarmont) compensator.

The object is placed at ±45° between crossed polars. Between the object and analyser a quarter-wave plate is inserted with its slow axis parallel or orthogonal to the transmission axis of the polarizer, and the analyser is rotated until the object appears maximally dark. If A is the angle in degrees through which the analyser has been rotated, the retardation Δ of the object is 2A in degrees or A/180 expressed as a fraction of a wavelength.

There is often some ambiguity about the results obtained using the quarter-wave method, and some care is always necessary. Firstly, in common with other methods using monochromatic illumination, an object with a retardation of (say) 0.2 wavelengths will give the same apparent result as one with a retardation of 1.2 or 2.2 wavelengths. Unless one is familiar with the sort of object being studied, it is therefore advisable to use first a less precise method such as the quartz wedge with white light, in order to measure the approximate retardation in whole wavelengths. Secondly, it is not possible with the quarter-wave method alone to distinguish between a retardation and an advance in the polarized light; thus a retardation of 0.2 wavelengths can easily be confused with an advance of 0.8 wavelengths. This can be particularly troublesome if the retardation is in the neighbourhood of half a wave-length.

Suppose an object is both birefringent and dichroic; with a retardation of Δ degrees and a ratio of the amplitudes of the light components leaving the object of tan Y. It is possible to analyse the emergent light completely using a rotatable quarter-wave plate and analyser (Halli-mond, 1953 p. 129; Goldstein, 1970a). The compensator and the analyser are rotated simultaneously or alternately until positions are found giving maximum extinction of the object. If the angle through which the compensator has been rotated is X (measured from the initial position of its slow axis parallel to the transmission axis of the polarizer) and the angle of the analyser (relative to the *new* position of the compensator) is θ, then

$$\tan(\Delta) = \frac{\tan(2\Theta)}{\cos(2X)} \qquad (14.1)$$

and

$$\tan(2Y) = \frac{\tan(90° - 2X)}{\cos(\Delta)} \qquad (14.2)$$

Neither the adjustments to the microscope nor the calculations are very easy with this method – a better method using *two* quarter-wave plates is described below.

14.5.8 Traditional analysis of dichroism with a rotating polar. The traditional but rather insensitive method of detecting the presence and sign of dichroism with a single rotating polar has been described above. If there is no change in the apparent intensity of the object as the analyser is rotated, it can be concluded that dichroism is absent (or slight). To obtain the ratio of the amplitudes of the components of light leaving the specimen parallel and orthogonal to the principal axis of the specimen, the object is placed at $\pm 45°$ between crossed polars. The analyser is now rotated through an angle A until the object appears maximally dark. The ratio of the amplitudes of the two components is given by ATN (RAD (45−A)) where A is less than 45°. The ratio of the corresponding intensities is (ATN (RAD (45−A))) ^ 2. Note that the measured ratio is not the same as the dichroic ratio – the latter is defined as the ratio of the absorbances parallel and orthogonal to the principal axes of the object, and requires the measure-ment of the separate transmissions and their conversion to absorbances.

14.5.9 Analysis of both birefringence and dichroism using two quarter-wave plate compensators. This method (Goldstein, 1970a) has the advantage that the calculations are relatively simple. In addition, all adjustments can be made using a Nakamura half-wave plate so that the method is capable of high precision. Between the specimen (at $\pm 45°$ to the polarizer) and the rotatable analyser are two quarter-wave plates. That nearer the specimen is fixed with its slow axis at $\pm 90°$ to the polarizer, while that nearer the analyser is rotatable. Assume the retardation of the light leaving the slow axis of the object is Δ relative to that leaving the fast axis, and the amplitude ratio of the components of light leaving the specimen is tan(Y), where Y is less than 45°. At compensation the azimuth of the rotatable compensator is then $\Delta/2$ relative to its initial position orthogonal to the polarizer, and the azimuth of the analyser is at (45° $-$ Y) relative to the new azimuth of the rotatable compensator.

The full procedure is as follows, assuming the polars are already accurately crossed, the object is at $+45°$, and the quarter-wave plate immediately above the objective is fixed accurately parallel to the polarizer (all these adjustments, and the most critical adjustments in the following list, are best made with the aid of a Nakamura plate). The second quarter-wave plate is initially withdrawn.

(i)   Rotate the analyser to bring the object to as near extinction as possible (check with Nakamura). The precision of the setting can be improved with a dichroic object (without systematically affecting the final result) by temporary slight rotation of the polarizer – anti-clockwise rotation if the direction of the maximum absorption of the object is at $+45°$, and clockwise rotation if the object is at $-45°$. Record the azimuth of the analyser as angle A.

(ii) Insert the second (rotatable) quarter-wave plate with its fast axis at angle A, i.e. parallel to the *new* position of the analyser.

(iii) Again adjust the analyser to give extinction of the specimen (check with Nakamura), and record its azimuth as angle B.

(iv) Calculate retardation $\Delta = A/2$, and amplitude ratio $Y = 45° -$ (B $-$ A). For a full discussion and proof of the method, consult the original article.

14.6 SUGGESTED EXERCISES USING THE NICOL PROGRAM

14.6.1 Rotation of analyser (fixed polarizer).   Withdraw all components except the polarizer and analyser from the optical path. Rotate the analyser either manually (Main Menu option 4) or automatically (Main Menu option 5). Note the change in intensity as the analyser azimuth changes. Where are the maximum and minimum intensities found?

14.6.2   Rotation of object between crossed polars.   With the analyser in the crossed position relative to the polarizer, insert the object into the optical path. Check (Main Menu option 1) that the object is birefringent but not dichroic, and if necessary modify the object parameters (Main Menu option 3). Rotate the object manually and/or automatically, and note the changes in intensity with object azimuth. At what azimuths are the maximum and minimum image intensities found?

14.6.3   Rotation of analyser with fixed polarizer and fixed object.   With the object at $+45°$ between crossed polars, rotate the analyser until the object appears with minimum intensity. Compare the intensities of the object and background at various analyser azimuths. At what analyser azimuths are (a) the minimum background intensity, and (b) the minimum object intensity found?

14.6.4   Rotation of dichroic object between crossed polars.   Redefine the object so that it is dichroic (e.g. transmission along one axis is 1.0 and along the other 0.5). Rotate the object between crossed polars. At what azimuth are the minimum and maximum object intensities found? With the object at $±45°$, rotate the analyser. At what azimuths are the maximum and minimum object intensities found? Compare the results with those of Exercise 14.6.2 and try to explain your results.

14.6.5   Birefringent object and quarter-wave plate.   Define the object to have a moderate birefringence (retardation of say $30°$), and make sure it is at $±45°$. Between the object and the analyser insert a compensator with a retardation of $90°$ with its slow axis parallel to the transmission axis of the polarizer (i.e. at $0°$), and rotate the analyser manually and/or automatically. At what azimuth is the minimum object intensity found? Can you relate this azimuth to the (known) retardation of the object? This exercise is a demonstration of the Sénarmont method.

14.6.6   Rotation of compensator (not quarter-wave!).   With the same object as before, at the same azimuth (i.e. at $±45°$), re-define the compensator to have a retardation of (say) $35°$. Rotate the compensator with the analyser in the crossed position relative to the polarizer, and note the azimuth of the compensator which gives the minimum object intensity. Try to relate this azimuth to the retardation of the object (use the Calculator to try out various expressions). This is a test of the Brace–Köhler compensator method.

14.6.7   First-order red (sensitive-tint) compensator.   Insert a full-wavelength retardation plate (i.e. a sensitive-tint or first-order red compensator – this is the initial or default state of the upper compensator in the system), oriented at $\pm 45°$ relative to the crossed polars. Choose the white-light option in the Main Menu, and choose to rotate either the compensator or the object. Note the colours of the background and object, and how they change as the object is rotated relative to the compensator. When the object and compensator are in 'addition' position, i.e. with their slow axes parallel, the retardation of the combination is raised so that the colour of the object is different from that when the object and compensator are in 'subtraction' position. With a slightly retarding specimen the 'addition' and 'subtraction' colours are typically blue and yellow respectively. This procedure enables the sign of the birefringence of the object to the ascertained.

14.6.8 Use of quartz wedge (white or monochromatic light). With all compensators withdrawn from the optical path, and the object at $\pm 45°$ between crossed polars, choose one of the quartz wedge options (with either white light or monochromatic green light illumination) from the Main Menu. Choose to rotate either the object or the quartz wedge. Note the bands where the wedge covers the background, and how the bands are shifted where the wedge covers the object. This is a rough method of measuring the retardation of the object, and also provides a way to find out the sign of the birefringence (find out how the wedge behaves on a known object before you try to use it on an unknown one).

14.6.9 Analysis of object of unknown birefringence. When you feel that you have a reasonable understanding of the methods so far used, make the retardation of the object random (Main Menu option 8), but restrict the possible retardation to a small value (you are prompted how to do this by the program) and leave the azimuth at $45°$ and both transmissions at 1.0 (i.e. no dichroism). Test for the presence and sign of the birefringence using a compensator of about $10°$ retardation, inserted either parallel or at right angles to an axis of the object. Alternatively, use the first-order red compensator (which retards green light by precisely one wavelength) with white light. Then try to measure the object retardation using either the Sénarmont or the Brace–Köhler method.

14.6.10 Analysis of object of unknown dichroism. With a non-birefringent object at $45°$ between crossed polars, make the dichroism random (leaving the azimuth and birefringence unaltered). Try to ascertain the sign of the dichroism by rotating the polarizer (no compensator or analyser in the optical path), or by *slightly* rotating the analyser clockwise or anti-clockwise from the crossed position (still with no compensator in the system). Using the Nakamura plate between crossed polars is another method for doing this. Then rotate the analyser either manually or automatically until the intensity of the object is at a minimum. What is this minimum, and at what azimuth is it found? Can you relate the azimuth to the dichroism of the specimen (find this out by again choosing Main Menu option 6, or touching function key F5, and typing Y when asked 'Do you want to give up now? (Y/N)'. Check your answer by using the Calculator).

14.6.11 Analysis of randomized object (unknown orientation, birefringence, dichroism). See if you can analyse a completely unknown object (randomized azimuth, birefringence and dichroism). If you get an even *approximately* correct answer, well done!

Chapter 15

How the Zernike Program Works

This chapter (see also Goldstein, 1991a, 1992) is intended for readers who have an interest in computing or who may wish to adapt the program for their own purposes. It is *not* essential reading if you simply want a better understanding of light microscopy!

15.1 MATHEMATICAL REPRESENTATION OF AMPLITUDE AND PHASE, AND THE COMPUTATION OF INTERFERENCE

To calculate what happens when two coherent light rays interfere with each another, one could plot sinusoidal graphs representing the amplitudes and relative phase of the rays, and by adding the amplitudes at point after point (taking account of negative and positive signs where necessary) obtain the amplitude of the product, or 'resultant'. This procedure is prohibitively laborious and time-consuming. Fortunately, the same result can be achieved by a *vector* method in which the amplitude of each ray is represented by the length of a vector, and the phase difference between the rays is proportional to the angle which the vectors make with each other. The resultant vector is then obtained using the parallelogram of forces (Figure 15.1). If vectors OA and OB represent respectively rays A and B, which differ in phase by the angle BOA, the phase and amplitude of the resultant ray R is given by the direction and length of vector OR. By simple trigonometry (the cosine rule),

$$OR^2 = OA^2 + OB^2 + 2.OA.OB.\cos(BOA) \qquad (15.1)$$

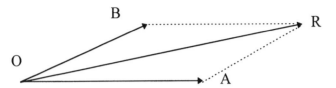

Figure 15.1   The cosine rule

and

$$\sin(ROA) = \frac{OB.\sin(BOA)}{OR} \tag{15.2}$$

Another way of computing the resultant of interference of two or more waves is often more convenient (Figure 15.2). The amplitude and phase of each ray is stored in the form of two numbers which represent the amplitudes of two hypothetical waves, a quarter of a wavelength out of phase with each other, which would on interference give the ray in question. One of these hypothetical waves is traditionally called the 'real' and the other the 'imaginary' component of the actual light wave in question. If a wave OR of unit amplitude is retarded by ROA degrees relative to a reference wave OA, the real and imaginary components of OR respectively have amplitudes OA = OR.cos(ROA) and OB = OR.sin(ROB).

The following are some special cases. If the wave OA is exactly in phase with the reference wave, its amplitudes of its real and imaginary components are respectively unity and zero. If OA is retarded by 90° (one quarter of a wavelength) relative to the reference wave, its real component is zero and its imaginary component has unit amplitude, while a wave retarded or advanced by half a wavelength has real and imaginary components with amplitudes of −1 and zero respectively.

This notation has a major advantage. To work out the resultant of the interference of any number of coherent waves it is necessary only to add all the real components and separately add all the imaginary components, in each case taking account of the signs. The two sums give us the real and imaginary components A and P of the resultant, the amplitude and phase of which can now readily be calculated – the phase angle equals $\arctan(P/A)$, and by the theorem of Pythagoras the amplitude equals the square root of the sum of the squares of A and P.

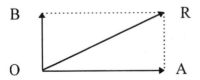

Figure 15.2   Real and imaginary components

In the program the arrays A(n) and P(n) hold respectively the real and imaginary components of the light rays at different points across the field. For example, the number held in 'box' A(5) represents the real component of the light wave at the fifth point from the left across the field, and similarly P(64) is the imaginary component of the light 64 places from the left of the field. The optic axis is in the middle of the field, which for the sake of symmetry ought (in the case of an array 64 units wide) to be at position 32.5. No half-positions are however possible, so the axis is taken to be at position 33.

Some mathematical manipulations are particularly easy with this notation. If we want for example to change the phase of the light at a given point (say point 5) by exactly one quarter of a wavelength, it is necessary only to change the sign of the value held in P(5), and swap it with the value held in A(5) using a temporary variable as a store:

LET temp = A(5)
LET A(5) = −P(5)
LET P(5) = temp

Changing the phase by half a wavelength is even easier − simply change the signs of both A(n) and P(n):

LET A(5) = −A(5)
LET P(5) = −P(5)

The function called FNA in the program works out the phase angle (in radians) of a ray from its real and imaginary components, and allows for the possibility that the angle may be greater than 90° ($\pi/2$ radians). A version of the algorithm can be found in lines 10050–10080 of the listing for the 'Slow Fourier Transform', below. Function FNR calculates the intensity of the light at a given point, i.e. it is the sum of the squares of the two components at that point.

15.2 THE FOURIER TRANSFORM

According to the Abbe theory of microscopical imaging, the pattern of light visible in the back focal plane of the microscope objective is caused by the interference of rays of light diffracted from points in the object. Rays coming from different points in the diffraction pattern in turn interfere to cause the real image visible at the front focal plane of the microscope eyepiece. The Fourier transform is a mathematical procedure which can be used to calculate either the diffraction pattern corresponding to a given object, or the image corresponding to a given diffraction pattern. Two successive applications of the Fourier transform in principle restore a set of data to their original form, but reversed left-to-right; remember that a real image formed by a single lens is also a mirror image. Four successive applications of the Fourier transform give a correct, erect image.

The fast Fourier transform (FFT) used in the Zernike program is an extremely rapid and efficient one-dimensional transform which manipulates data corresponding to a cross-section through the microscopic field. In order to handle an image as seen 'face-on' a two-dimensional Fourier transform would be needed, but such a transform is relatively very slow and requires large amounts of computer memory, and would only be practical with a very powerful computer.

It is almost impossible for the mathematically uninitiated to work out how an FFT works simply by inspecting its computer coding. Some insight can however be obtained by studying the following BASIC procedure, called PROCSFT. Although this 'Slow Fourier Transform' is much slower and needs much more memory than the FFT, it performs precisely the same function and is relatively easy to understand.

PROCSFT starts with the amplitude and phase of the rays at each point in a plane (say the object plane) held in arrays A(n) and P(n). Arrays A2(n) and P2(n) are used to hold intermediate results, and end up holding the amplitude and phase of the rays in the back focal plane of the objective. In the listing J% refers to points in arrays A2(n) and P2(n), and I% to points in arrays A(n) and P(n). For each point in A2(n) and P2(n) the values held are initially set to zero (line 10020). Then the contribution of energy to that point is calculated for every point in the first pair of arrays: in line 10050 the amplitude ('amp') of the light at a point in A(n) and P(n) is calculated as the square root of the sum of the squares of the real and imaginary components, and in lines 10060–10080 the phase ('phase') of the light when it leaves the object plane is calculated (line 10080) using the same algorithm as in function FNA. The phase ('new-phase') of the light when it arrives at the second plane is related (line 10090) to both the old phase and the distances of points I% and J% from the optic axis. For example, if either point is on the optic axis no phase change is introduced. The real and imaginary components of the light in the second plane respectively equal the product of the amplitude with the sine and with the cosine of the phase (lines 10100 and 10110), and these are added into arrays A2(n) and P2(n). Finally, when this has been done for every point in arrays A(n) and P(n), the sums held in A2(n) and P2(n) are divided (line 10140) by the square root of N in order to maintain a constant total intensity of light, and copied back to the original arrays A(n) and P(n).

```
10000 DEF PROCSFT:REM A 'slow' Fourier transform
10010 FOR J% = 1 TO N:REM points in the arrays A2 and P2
10020 LET A2(J%) = 0:LET P2(J%) = 0:REM set amplitude and phase
      to zero
```

```
10030 FOR I% = 1 TO N:REM points in arrays A and P
10040 LET    amp = SQR(A(I%)*A(I%) + P(I%)*P(I%)):REM    original
      amplitude
10050 LET A = A(I%)
10055 LET P = P(I%):REM temporary values
10060 IF ABS(A) < 1E-9 THEN LET A = 1E-9:REM to ensure no divi-
      sion by zero
10070 IF ABS(P) < 1E-9 THEN LET P = 0
10080 LET phase = ATN(P/A):REM original phase
10085 IF A < 0 THEN LET phase = phase + PI:REM in case 'phase'
      more than PI
10090 LET newphase = phase + (I%-axis)*(J%-axis)*2*PI/ N
10100 LET A2(J%) = A2(J%) + amp*COS(newphase)
10110 LET P2(J%) = P2(J%) + amp*SIN(newphase)
10120 NEXT I%
10130 NEXT J%
10140 FOR I% = 1 TO N

10145 LET A(I%) = A2(I%)/SQR(N)
10146 LET P(I%) = P2(I%)/SQR(N)
10147 NEXT I%:REM data back to original array and normalize for size
      of array
10150 ENDPROC
```

Because of the use of two 'nested' loops in this 'slow' Fourier transform, the number of calculations and therefore the time taken is proportional to the square of the number of points (N) in the arrays. It can be shown that the time taken by a fast Fourier transform is approximately proportional to N multiplied by log(N); this is much faster, particularly with large values of N.

15.3 REPRESENTATION OF FINITE OBJECTIVE NA

Two consecutive applications of the Fourier transform to a set of data return the set to its original form (apart from left-to-right inversion). This corresponds to a hypothetical situation in which the angular semi-aperture of a (dry) objective equals 90°, i.e. the objective admits all the diffracted light emitted by the object and the image perfectly represents the object. If the objective NA is less than unity, as must be the case in real life, the highest orders of diffracted light may be eliminated in which case the image will be to some extent imperfect. The program simulates a finite objective aperture by setting to zero the values held in the extremes of the arrays representing the diffraction pattern in the back focal plane of the objective. For example, if each array has 64 elements and the defined NA is 0.5, array elements from n = 1 to n = 16 and from n = 50 to n = 64 are set to zero; the values at n = 17 and n = 49, corresponding to the edges of the aperture stop, are divided by the square root of 2, which is equivalent to halving the intensities at these points. Application of the Fourier transform to the truncated sets of data calculates the corresponding imperfect image.

15.4 FOCUSSING

Zernike (1942) pointed out that when a microscopic object is in focus the optical distance from a given point in an object to the corresponding point in a perfectly focussed image is constant, irrespective of the path taken by the light, and all points in the Fraunhofer diffraction pattern are the same distance from the axial point in the image plane. Throwing the object out of focus is equivalent to introducing a phase difference between the different beams of light reaching a given image point from the Fraunhofer diffraction pattern, and if the object is an extended strip (as in the present program) the phase differences in the diffraction pattern are a function of the square of the distance from the axis. The following construction (following a suggestion by Barer, 1986) shows that throwing an object z wavelengths out of focus introduces a phase difference of $z(SQR(1 - NA^2))$ between the axial direct light and the diffracted light at a position in the aperture corresponding to NA.

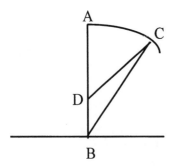

Figure 15.3 Diffraction pattern – change in focus

In this diagram the optic axis is AB, and the change of focus z at the centre point B of the object is BD. At a point C on the spherical wavefront of the light emerging from the object, the distance from point D is CD, which by the cosine rule is given by

$$CD^2 = BD^2 + CB^2 - 2.BD.CB.\cos(DBC)$$

Now CB equals AB, which is a relatively large distance F. The change in focus at point C is therefore

$$F - \sqrt{F^2 + Z^2 - 2F.Z\sqrt{1 - NA^2}}$$

which can be shown to be equivalent to

$$Z\sqrt{1 - NA^2}$$

if F is much larger than Z. In the program the object is assumed to be infinitely elongated perpendicular to the screen. The program therefore simulates incorrect focus by systematically introducing a phase difference of Z.SQR(1 − NA^2) into the diffraction pattern, where Z is the deviation from best focus at the optic axis measured in wavelengths, and NA is the distance of the point from the optic axis, measured in terms of numerical aperture.

If one wishes to compute the (Fresnel) diffraction pattern at an arbitrary distance from the object this can be done by performing a FFT on the light leaving the object, to compute the Fraunhofer diffraction pattern. The phases of this pattern are then altered systematically to change the focus by any desired amount, and another FFT is performed without any restriction on the aperture of the objective. This computes a perfect but out-of-focus image of the object, which represents the corresponding Fresnel diffraction pattern in the object space.

15.5 SPHERICAL ABERRATION

Martin (1966, p. 113 *et seq.*) discusses monochromatic aberrations in terms of the deviation of wavefronts from sphericity. In primary spherical aberration, sometimes called the 'first Seidel aberration' or S1, the deviation is proportional to the fourth power of the distance from the optic axis. If the focus error and the primary spherical aberration are opposite in sign they may partially cancel each other out, so that in the presence of primary spherical aberration the 'best' image (i.e. the image which most closely resembles the object) is in general obtained by throwing the object slightly out of focus. If only the zero and first orders of diffracted light are present in the diffraction pattern, as in the case of a diffraction grating with a small interval, it may be possible to obtain a perfect image in this way. In general, however, complete correction of spherical aberration cannot be obtained, because even if at some distance from the optic axis the focus error and the spherical aberration are equal in magnitude and opposite in sign, nearer the optic axis the focus error will predominate and further from the optic axis the spherical aberration will have more effect.

The following algorithm modifies the diffraction pattern according to the deviation from best focus Z and the primary spherical aberration S1, both measured in wavelengths:

```
REM a% = N/2 + 1 is the position of the optic axis
REM FNA(I%) returns the phase angle at point I%
REM FNR(I%) returns intensity at point I%
FOR I% = 1 TO N:REM points in the diffraction pattern
LET NA = ABS(a%-I%)*2/N
LET tempamp = SQR(FNR(I%))
LET tempphase = FNA(I%)
LET tempphase = tempphase + 2*PI*Z*SQR(1-NA^2):REM focus
LET tempphase = tempphase +2*PI*S1*NA^4:REM aberration
LET A(I%) = tempamp*COS(tempphase)
LET P(I%) = tempamp*SIN(tempphase)
NEXT I%
```

15.6 PARTIALLY COHERENT ILLUMINATION

There are several alternative ways of computing this. The method originally described (Goldstein, 1991a) involved three nested loops, but the faster current algorithm takes advantage of the fact that in partially coherent illumination neither the focus nor the aberrations of the condenser affect the final result. Köhler illumination is assumed, with the lamp filament imaged into the front focal plane of the condenser. The algorithm starts by performing a FFT on the data representing light from the object, held in the arrays from point 1 to point N; the computed diffraction pattern is copied into a convenient higher part of the arrays, thus freeing the lower part of the arrays for other uses.

The position of the centre of the Fraunhofer diffraction pattern, relative to the optic axis in the back focal plane of the objective, depends on the obliquity of the incident ray of light under consideration – the more oblique, the further is the diffraction pattern displaced laterally. A separate calculation is therefore carried out for light from each (mutually incoherent) point K% in the front focal plane of the condenser – since the condenser aperture is less than unity, fewer than N points need be considered. For each point I% in the object's diffraction pattern, the distance D% from the optic axis a% (equal to N/2 + 1) is calculated as follows:

```
LET D% = a% − K% + I%
IF D% < N THEN LET D% = D% − N:REM wrap-around
IF D% > 1 THEN LET D% = D% + N:REM wrap-around.
```

The values in the arrays at point I% are then copied to the lower part of the arrays, to the calculated point D%. When this is completed for all the points in the diffraction pattern, the copied diffraction pattern is appropriately truncated to allow for the finite objective aperture, and if necessary its phases are altered as already described to simulate imperfect focus and/or spherical aberration. Performance of the FFT now computes the corresponding image; the intensity at each point is added to a temporary store elsewhere in the arrays, where the final image is accumulated. When all points K% have been thus dealt with, the final accumulated image is transferred to the lowest part of the arrays for display or examination. The whole process, although quite time-consuming, involves only two nested loops and therefore is considerably quicker than the original method.

Computations for complete coherence, conventional fluorescence and confocal scanning are in principle similar to that for partial coherence, and also use only two nested loops. The calculations within each loop are rather shorter than in the case of partial coherence, and the whole procedures are therefore somewhat faster.

15.7 SOME INHERENT LIMITATIONS OF THE PROGRAM

(1) The one-dimensional fast Fourier transform used in the program manipulates data corresponding to a cross-section through the microscopic field. To handle an image as seen 'face-on' would require a two-dimensional Fourier transform, which would only be practical with a very fast and powerful computer.

(2) Objects, images and the objective aperture should be regarded as being one-dimensional, with a finite extent laterally but indefinitely long in a plane perpendicular to the field of view. For example, what appears on the screen to be a pin-hole actually represents a cross-section through a slit. The distribution of energy in the diffraction-limited image of a point (the Airy disc, see e.g. Fig. 2.38 of Martin, 1966) is slightly different from that of a slit as computed in the present program. The objective aperture being one-dimensional, the computed intensity of the image of e.g. a fluorescent object is directly proportional to the objective aperture and not, as in the two-dimensional case, to the aperture squared.

(3) Similarly, throwing an object out of focus introduces phase differences in the Fraunhofer diffraction pattern proportional to the square of the distance of each point in the pattern from the optic axis. If the object were a discrete point or disc and the aperture of the lens round, the areas of each zone in the back focal plane of the objective, corresponding to a particular small range of phase differences, would be equal (Martin, 1966, p. 91–93), so therefore would be the intensity of the light reaching the image from each

member of such a zone pair. Martin points out that with a change of focus of (say) one wavelength, each zone in the back focal plane can be matched with a zone of equal area, but half a wavelength out of phase – the light from each such pair interferes destructively in the centre of the image of a point object, which therefore appears black. In the one-dimensional computer model, however, the intensities of each member of a pair are *not* equal, and complete destructive interference does not occur under the same circumstances. The effects of de-focussing will accordingly be quantitatively different in the model.

(4) Despite the fact that the object is in general inherently one-dimensional, with negligible thickness in the z axis, when using axial coherent illumination it is possible to simulate an object consisting of a number of discrete planes, separated by arbitrary distances (see Section 16.8, p. 139).

(5) In some textbooks of optics distances in the image plane are expressed in *optical z units*, the unit being defined as a wavelength of light divided by the NA of the objective used. The unit of distance across the microscopic field in the program is half a wavelength of light. With an objective aperture of (say) 0.1, one wavelength in the object or image plane in the program then corresponds to 0.1 'optical units'. No intervals smaller than a half-wavelength can be simulated directly in the program, and the widths of objects and the distances between them must be whole multiples of the basic unit. If it is assumed that the effects of diffraction are linearly related to the objective NA, it is possible (with some reservations!) to simulate the imaging of an object with smaller intervals between its points, by an objective of given NA, by defining a wider object and a smaller NA. For example, the imaging of an object 5 half-wavelengths wide by an objective of NA 0.5 is similar to the imaging of an object 10 half-wavelengths wide by an objective of NA 0.25; the latter case could be interpreted as the imaging of an object with a quarter-wavelength interval between its points by an objective of NA 0.5.

(6) The simulated ideal objective is normally aberration-free, but primary spherical aberration can be specified. The lens has a maximum angular aperture of 180°, corresponding to a 'dry' lens with an NA of 1.0 (which is not in fact physically possible). To simulate an immersion lens with an aperture greater than unity, simply regard the objective NA in the program as representing the sine of the admission half-angle of the lens, and not (as usual) the sine multiplied by the refractive index of the immersion medium. The unit of distance in the object and image planes is now a wavelength of light as measured in the immersion medium, and not (as usual) in vacuum. To illustrate this, suppose that an oil-

immersion lens has a numerical aperture of 1.14 when used with oil of refractive index 1.52. The lens has an angular aperture which is identical with that of a 'dry' lens with a numerical aperture of 1.14/ 1.52 = 0.75. To simulate the oil-immersion lens specify an aperture of 0.75, and interpret one wavelength shown on the computer screen as representing 1/1.52 = 0.66 of a wavelength measured in vacuum.

(7) If a Fourier transform procedure is applied to data representing light emerging from an object the corresponding diffraction pattern in the back focal plane of the lens is computed. Applying the Fourier procedure a second time, to the transformed data, computes an image inverted left-to-right. To obtain a correctly oriented image *two more* Fourier transformations are necessary, corresponding to the introduction of a second lens (e.g. an eyepiece) into the optical system. The Fourier algorithm used in the program has however been slightly modified by inverting the sign of the final 'imaginary' components. There is now no left-to-right image inversion after only two applications, but this is at the cost of a slightly incorrect representation of the phases of the light in the diffraction pattern. This is of little practical significance but ought to be remembered when studying such topics as the theory of phase contrast microscopy.

(8) Several algorithms used, including the FFT, employ 'wrap-around'. For example, in the simulation of oblique coherent illumination the whole Fraunhofer diffraction pattern is shifted laterally, and higher orders of diffracted light on one side of the optic axis may fall outside the limits of the objective aperture. Values which disappear off one side of the screen may however reappear on the other, and this physically unrealistic phenomenon should be taken into account in the interpretation of results. Another effect of 'wrap-around' (or perhaps another way of thinking about it) is that the defined object and the computed image represent slices through an infinitely extended plane. Thus if a single object is defined in the microscopic field, identical objects are implicitly created on both sides, in invisible parts of the object plane; under some conditions parts of the images of these 'invisible' objects may spread into and affect the visible computed image.

(9) The program assumes that the light is perfectly monochromatic, and ignores both polarization and the effect of obliquity on the intensity of diffracted rays.

(10) The algorithm used in the program to compute the effect of stray light (glare) in a lens retains pre-existing phase relationships in the image. In reality, after the multiple reflections responsible for glare the redistributed energy is effectively incoherent. The discrepancy

does not affect intensities in the final image assuming glare is due only to the objective, but may cause slight errors if glare is specified in a condenser of finite aperture.

(11) Microscopists often adjust the lamp intensity to compensate for alterations in image brightness due to such factors as a change in the numerical aperture of the condenser. The program attempts to do the same. This may occasionally lead to anomalous results, so that some caution should be used in comparing absolute image intensities obtained under different instrumental conditions.

(12) The program deals primarily with transmitted-light illumination, although one diagram accessible in the Zernike program illustrates a simple form of epi-illumination fluorescence. Most of the imaging modes illustrated in or simulated by the program, although involving transmitted light, can be regarded as applying also to epi-illumination, simply by imagining the ray diagrams to be 'folded' so that a single lens serves as both objective and condenser. This applies particularly to the fluorescence and confocal fluorescence modes.

Chapter 16

How to Use the Zernike Program

16.1 LOADING AND RUNNING THE PROGRAM

The compiled TurboBasic programs are distributed on a CD-ROM disc suitable for a IBM-compatible computer, ideally not older than a 486 model with 16-colour VGA graphics.

It is first necessary to copy the programs and accessory files from the CD-ROM to the hard disc. Turn on the computer and wait for the DOS prompt (if you are using WINDOWS, double-click on the DOS icon). While logged on to the c: drive, type

```
md zernike (ENTER)
cd zernike (ENTER)
copy d:*.* (ENTER)
```

A file called readme.txt may contain information too recent to be included in this book, and is worth reading (using any word processor) before proceeding.

Using Windows95 or Windows98 it is best to run the programs in a Window, by clicking on the file called run.bat using Program Manager or Windows Explorer. With an older machine using DOS, if you intend to send program data or diagrams to an attached printer, before running Zernike it may be necessary to load a printer driver (e.g. 'graphics' or 'graphics color1' – consult your DOS handbook about this). If you want to use the mouse when running from Windows 3.1 it may be necessary to load a 'mouse driver' before entering Windows; type MOUSE(ENTER) before typing WIN(ENTER) (Windows 95 and Windows 98 do not require this).

On running the Zernike program you are first given the choice of one of the four programs which constitute the 'Simulated Microscopy' set. These are the Zernike program itself, and the Kohler, Snellius and Nicol programs which are described elsewhere in this book. The following deals only with the Zernike program proper.

The Zernike program enables you to define an idealized microscopic specimen or to 'adjust the microscope', i.e. change various imaging parameters. The specimen may be one or more discrete objects on a uniform background, or a diffraction grating. In all cases object properties such as position, width, transmittance and retardation are specified. Specimens should be regarded as being indefinitely elongated in a plane perpendicular to both the optic axis and the VDU, but in general of negligible geometrical thickness in a direction parallel to the optic axis (see Section 16.8 for how a specimen consisting of two or more layers can be simulated). The Main Menu allows you to change the transmitted illumination from coherent, axial transmitted light to oblique coherent light, partial coherence (finite condenser aperture), or complete incoherence. The imaging options then available include bright-field, phase-contrast, central and peripheral dark-field, Schlieren, Gegenfeld, modulation contrast and apodization. Alternative modes of microscopy accessible from the Main Menu are interference contrast, fluorescence, and confocal scanning (both fluorescent and transmitted). The width of the microscopic field can be altered giving a compromise between image detail and speed of computation, and when studying glare (stray light) the width of field illuminated may be reduced to simulate the effect of a field stop.

In the next stage of the program the properties of the light passing through each part of the microscopic field (specimen and empty background) are graphically shown on the computer screen; the abscissa represents distances across the microscopic field measured in wavelengths of light, with the optic axis in the middle of the screen. The ordinate in the upper part of the display can be amplitude, absolute amplitude (i.e. ignoring signs) or intensity (amplitude squared). Values are scaled to fill the available space. The lower part of the screen displays the relative phases of the light in different parts of the field. Phase retardations and advances are both represented as retardations, for example an advance of 0.1 wavelength is shown as a retardation of 0.9 wavelength.

If coherent transmitted illumination has been chosen, the Fraunhofer diffraction pattern at the back focal plane of the microscope objective is computed using a fast Fourier transform (FFT), and the results are displayed with amplitudes (or intensities) and phases as before; the abscissa is now however labelled linearly in terms of objective numerical aperture (NA), not wavelengths. With transmitted coherent, partially coherent and incoherent illumination the mode of microscopy (e.g. bright-field, dark-field, phase-contrast, etc.) is now chosen, and parameters such as the objective NA and plane of focus are specified. The appearance of the back of the objective, as modified by the defined imaging system, is then displayed (together with the modified Fraunhofer pattern in the case of coherent illumination), and the final image is computed and displayed. If fluorescence or confocal scanning have been specified in the Main Menu, you are asked as before to define various parameters of the objective, and possibly also of the condenser. The amplitude (or intensity) and phase of the light passing through the object and in the corresponding image are displayed as with coherent illumination, but no Fraunhofer diffraction pattern is shown.

The cycle can be repeated with or without alteration of the object or the imaging system, which makes it possible for the user to acquire an almost intuitive feeling for microscopic image formation. Some theoretical knowledge (as provided in an elementary way in this handbook) is however a help in deriving maximum benefit from the program.

16.2 ENTERING INFORMATION

At various points in the program you are requested to make choices or provide data.

In answering a question such as

```
PRINT THIS SCREEN? (Y/N)
```

briefly press key Y or N to represent 'yes' or 'no' respectively.

If a numbered list of options on the screen is followed by an instruction such as

```
TOUCH CHOICE 1 to 7
```

simply press the chosen digit key or one of the function keys (the actions of these are described below). If a mouse is present it is sometimes possible to select a choice by placing the mouse pointer on the option (which then lights up) and clicking the mouse button.

A typical instruction is

```
Number of objects "Nob%" (1 to 5) = 2
Press RETURN to accept old value
OR type new value and press RETURN
```

In this example *Number of objects* describes the variable to be defined, *Nob%* is the symbol used in the program for this variable, 1 and 5 are respectively the minimum and maximum values allowed, and 2 is the current value. Pressing ENTER (labelled RETURN on some machines) without first typing in a number accepts the current ('default') value. Clicking a mouse button with the pointer over an empty part of the screen is equivalent to pressing ENTER. Variables are given reasonable values at the start of the program, and whenever a variable is altered its new value becomes the default value.

To change a variable, type a number or expression on the keyboard, using a full-stop to represent the decimal point and the DELETE key to correct any mistake. Then press RETURN. The computer accepts explicit numbers such as 1.5 or -3, and expressions such as 3/4, TAN(OPD(2)/PI) or 0.5*BOPD. In these examples /, PI and * respectively represent the division sign, π and the multiplication sign. OPD(3) and BOPD(1) represent the current values of variables used in the program to describe properties of the objects (see Table 16-2). For example, to make the retardation of the third discrete object twice its previous value simply type 2*OPD(3) when asked to enter the new value.

16.3   ERRORS AND ERROR MESSAGES

Z is a recognized program variable, representing the current deviation from best focus of the objective. 5*Z and Z/2 are therefore valid expressions for inputting data, although the numbers they represent will be rejected if too large or too small. Examples of expressions which are not permissible are 1/0, 1/(N−N) and 2*m, the first two because they imply division by zero, and the last because no variable m is recognized by the program. If a mistake is made an explanatory message appears on the VDU, and after you press any key the program carries on.

When a graphics screen has been displayed a box appears containing 'buttons' which can be clicked with a mouse (if one is present) to access various functions. Alternatively, function keys (labelled F1, F2 etc.) can be pressed to obtain the same effects. Pressing the Space Bar or the Enter key (or clicking the left-hand mouse button with the mouse pointer anywhere on the screen except over a button) allows the program to proceed. Pressing the right-hand mouse button accesses the Help function.

16.4   ACTIONS OF FUNCTION KEYS

A number of special functions can be accessed by pressing a function key at various points in the program. If a computer mouse is present it may be possible to activate a function key by clicking the left mouse button when the pointer on the screen is over a labelled rectangle.

After using a Function key the main program resumes, more or less where it was interrupted.

Table 16-1   Function keys

Key	Action
F1	'HELP'. Information, more or less appropriate to the part of the program in use, is provided on the screen. When in the HELP mode it is possible to access any HELP page (including the HELP menu) before returning to the main program. It is possible to 'scroll' upwards or downwards in a given help section.
F2	'PEEK'. Display the contents of the numerical arrays $A(n)$ and $P(n)$, which hold the current values of intensity and phase in the object, diffraction pattern or image.
F3	'DISC'. Save current data to disc with an optional comment, read in saved data, or wipe some or all stored data from the disc. This facility is useful to store the results of lengthy computations.
F4	'FIGS'. Examine diagrams illustrating ray paths in microscope systems, or the theory of diffraction.
F5	' ≪ '. ('Fast return'). Return to the Main Menu.
F6	' < '. Return to the start of the current program section. This is useful if a mistake is made in entering a value.
F7	'PLOT'. Change the screen display mode. The modes affect the upper part of the display (intensity or amplitude). The main modes are 'smoothed' (adjacent data points are joined), 'steps' (data points are represented by a rectangular block, like a histogram), 'delta' (vertical lines indicate the value of each data point) and OFF. In each main mode the ordinate can be intensity, absolute amplitude, or signed amplitude. Touching F7 repeatedly 'toggles' through all the options in turn and then returns to the default option ('smoothed, density').
F8	Load and run the Kohler, Snellius or Nicol program.
F9	'.PCX'. Save the current graphics screen to a .PCX file on disc, which can later be used by a suitable graphics program. If running in DOS, the screen can be sent directly to a printer by simultaneously pressing SHIFT and PRINT-SCREEN; in Windows the screen can be sent to the Clipboard for later access by another Windows program, by pressing ALT and PRINT-SCREEN.
F10	'D.I.Y.'. Individual procedures and functions of the Zernike program can be accessed in any sequence. A 'Calculator' is included.

16.5   THE MAIN MENU

This appears at the start of the program proper. Options 0, 1 and 2 define the specimen, and other options alter properties of the microscope. Choose an option by touching a digit key, and follow the further information and/or instructions which appear on the VDU.

(0) *Recall previous object.* Available only after a specimen has already been fully defined.

(1) *Discrete object(s).* These may have a square, circular or prismatic (diamond-shaped) profile. Up to 5 specimens can be defined, each characterized by shape, width, distance from the optic axis, transmittance and retardation. If two or more objects are defined an object may either replace, or be added to a previously defined object. The

'replace' option can be used to define a self-luminous object, by giving the first object the full width of the field and a transmittance of zero, and the second object a narrower width and a finite transmittance. 'Adding' the second object can be used, for example, to simulate a spherical cell containing a spherical nucleus with a different refractive index.

OW(n) is the width of object no. n, and OF(1) the distance between the centre of object 1 and the optic axis, both measured in half-wavelength units. It should be noted that due to the discrete nature of the computation, an object is not continuous but consists of a number of discrete points. If the width of a given object is (say) three units, the distance from one edge of the object to the other is therefore not 1.5 but 1.0 wavelengths. If OF(n) is zero the object is central in the field, while with positive and negative values respectively the centre of the object lies to the right or the left of the optic axis. The transmittance T can have any value between zero (opacity) and 1.0 (complete transparency). OPD (minimum -9 wavelengths, maximum +9 wavelengths) is the retardation relative to the background; an object with an OPD of zero has the same refractive index as the surrounding medium. If the object has a 'square' profile, a (uniform) retardation of (say) 5.5 wavelengths is equivalent to 0.5 wavelengths, i.e. the whole wavelengths don't count. The defined retardation of a 'circular' or 'prismatic' object applies only to its centre, and elsewhere its retardation is some fraction of the maximum. Two 'circular' objects respectively with 0.5 and 5.5 wavelength retardation at their centres may therefore appear very different under identical imaging conditions.

(2) *Periodic objects (diffraction gratings)*. Gratings with 'square', 'sawtooth' or 'sinusoidal' profiles can be defined. The grating interval C% in half-wavelengths has minimum and maximum values respectively of one wavelength, and the full width of the field. BW% in half-wavelengths is the width of the bar of a square grating or the leading edge of a sawtooth grating. The peak grating absorbance BA has a maximum value of 9 absorbance units, corresponding to a transmittance of 0.0000001%; the maximum retardation BOPD is ± 9 wavelengths. A uniform gradient the full width of the field can be defined by specifying a sawtooth grating with an interval equal to the field width, and with a leading edge width of either zero or the full width of the field.

(3) *MODE*. This option toggles through normal transmitted-light microscopy, differential interference contrast, normal fluorescence microscopy, transmitted-light scanning, fluorescence scanning and microdensitometry. Scanning is by default 'confocal' and diffraction limited, but the objective and condenser can be given different planes

of focus, numerical apertures, spherical aberrations and scanning spot sizes.

(4) *Field width.* This can be as small as 4 or as large as 128 wavelengths. With relatively wide fields more detail is shown in the image, but the computer takes longer to calculate results. This is especially important with partly coherent illumination and a relatively slow, old computer.

(5) *Field stop.* The area of specimen illuminated is normally the full width of the microscopic field, but can be reduced, for example to demonstrate the effect of the width of the illuminated field on the degradation of the image by glare. With coherent illumination (see below) the field stop has perfectly sharp edges and is situated in the same plane as the object, but with other modes of illumination the imaging of the field stop into the object plane is affected by the defined properties of the microscope condenser.

(6) *Illumination.* Transmitted-light illumination can be coherent (zero numerical aperture NAc), incoherent (unity numerical aperture and zero focus, glare and spherical aberration errors) or partially coherent (numerical aperture greater than zero but less than unity, or with unity aperture and finite values of condenser focus, glare or spherical aberration). With coherent illumination, which can be axial or oblique, several imaging options including phase-contrast and dark-field are available in addition to bright-field.

16.6   THE IMAGING SYSTEM

The numerical aperture, focus, glare and spherical aberration of the objective are specified in all imaging options. In many options additional parameters must be defined.

Properties of the objective:

Numerical aperture NA. This is from zero to the initial default value of 1.0, inclusive. An entered number is rounded to the nearest permitted value. An 'ideal' objective with an NA of unity has an admission half-angle of 90° (impossible in practice!). Reducing the NA in general reduces the intensity and degrades the quality of the image.

Glare G. This is defined as the fraction of the intensity at any point in the object which is spread out over the whole image plane. It can have any value between zero and unity. Glare values of around 0.05 (5%), commonly found experimentally, are mainly due to multiple reflections at air-glass surfaces in the objective, and can degrade image contrast significantly.

Deviation Z from best focus, in wavelengths. With the default value of zero the specimen, assumed to be of negligible thickness, is precisely in focus. The minimum and maximum values of Z are respectively -999 and $+999$ wavelengths. Throwing the specimen out of focus in general blurs the image, but can increase the visibility of a transparent object when using coherent illumination, or to some extent correct an image impaired by spherical aberration.

Primary spherical aberration S1, in wavelengths. The default value is zero. Spherical aberration can vary from -999 to $+999$ wavelengths. Spherical aberration and incorrect focus affect the diffraction pattern differently; the effects of focus are proportional to the square of the distance from the optic axis while those of spherical aberration are proportional to the fourth power of the distance.

With transmitted (both coherent and partially coherent) illumination there are seven main imaging options, selected in the image menu:

(1) *Bright-field*. No parameters are defined apart from those of the objective. With an objective aperture of 1.0 and zero glare, focus error and spherical aberration the image is identical with the original object.

(2) *Phase contrast*. The specimen is assumed to be illuminated with coherent light, with a disc-like phase plate centrally situated in the back focal plane of the objective. An A-type phase plate reduces the intensity of the central (zero-order diffracted, or direct) light, while a B-type plate absorbs part of the diffracted light passing through the outer zone of the objective aperture. Definable parameters of the phase plate are the radius PW, transmission PT and retardation PR.

The default phase plate is an A-type plate which advances the direct light relative to the diffracted light by a quarter of a wavelength. This corresponds to an ordinary commercial phase objective, which gives 'dark' or 'positive' phase contrast with a moderately retarding specimen. 'Bright' or 'negative' phase contrast is obtained if the phase plate retards the direct light by a quarter-wavelength.

In 'ideal' phase-contrast the phase plate is of zero radius and only affects the phase of the axial (zero-order diffracted) light. In practice the radius PW of the phase plate (expressed as a fraction of the numerical aperture of the objective) is finite, and the plate affects some of the diffracted light as well as the direct light. The phase plate transmission PT has a minimum value of zero and a maximum value of unity. With small values the system is more sensitive to objects of low retardation but also more liable to give misleading results with more highly retarding specimens. If PT equals zero all the direct light is absorbed, and phase contrast in effect becomes central dark-field illumination (see below).

(3) *Central dark-field*. The intensity but not the phase of the light passing through the centre of the back focal plane of the objective is reduced

by an axial absorbing stop. The two parameters to be defined are the diameter and the transmittance of the stop. The diameter DW, expressed as a fraction of the diameter of the back of the objective, can have any value between zero and 0.75. An ideal, zero-radius stop affects only the axial (non-diffracted) light, while one of finite radius absorbs in addition some of the diffracted, more peripheral light. The stop transmittance DT has limits of zero and unity. Zero transmittance gives central dark-field (central dark-ground) illumination, which shows transparent objects with their edges emphasized.

(4) *Schlieren and oblique dark-field*. The diffracted light on one side of the optic axis is completely absorbed, as if by a knife-edge in the back focal plane of the objective. With Schlieren the intensity of the zero-order diffracted light is reduced by 50%, while with peripheral dark-field the central light is completely removed together with a varying proportion of the diffracted light on the opposite side of the optic axis (depending on the diameter specified for the central stop). The opaque central stop is therefore given zero radius for Schlieren, and a finite radius and zero transmittance for peripheral dark-field. Schlieren images are asymmetrical, with edges and gradients darker on one side and brighter on the other; dark-field images are symmetrical. Oblique dark-field can alternatively be simulated by specifying obliquity of the coherent illumination in the Main Menu, and then in Bright-Field mode choosing an objective NA small enough to cut off the (displaced) direct light.

(5) *Gegenfeld*. The phase of the diffracted light on one side of the objective aperture is changed by half a wavelength, and the transmittance Gt of the zero-order light may be given any value between zero and unity. If Gt = 0 the effect is similar to dark-field but with even more striking contrast at specimen edges, while if all the zero order light is transmitted (Gt = 1) the imaging is asymmetrical and somewhat resembles Schlieren or differential interference contrast. Other values of Gt give intermediate results.

(6) *Hoffman's 'Modulation Contrast'*. A 'modulator' at the back focal plane of the objective has three zones which respectively transmit 100%, 15% and 1% of the light. The offset HO of the modulator from the midline determines the position of the narrow 15% zone relative to the optic axis. If this zone is placed at the edge of the objective aperture, with an equivalent obliquity of the coherent illumination, an effect is obtained similar to that given by some forms of apodization. The default value of HO is the current value of the condenser obliquity.

(7) *Apodization*. The results obtained depend in a complicated way on the maximum absorbance and maximum retardation allowed, and the apodization function specified. The defined apodization function relates the thickness of the plate in the back focal plane of the

objective to the distance X between the optic axis and a given point in the back focal plane. For example, if the function is simply X, the thickness of the apodization plate (and hence both the absorbance and the retardation) increases linearly with distance from the optic axis. The opposite effect is obtained if the function is $1 - X$. If the function is $X < 0.25$ the apodization plate has a uniform effect at numerical apertures between zero and 0.25, and none at greater apertures. A huge variety of effects can be obtained by suitable choice of function; in each case the effect is graphically shown on the VDU together with the altered diffraction pattern. The apodization function can be either symmetrical or asymmetrical about the optic axis.

The following imaging modes in addition to transmitted illumination, are selected using option 3 in the Main Menu.

Interference contrast
The appearance of the image depends on the shear distance, the distribution of energy between the reference and signal beams (normally equal), and the relative retardation between the signal and reference beams (normally a quarter-wavelength). Differential (Nomarski) interference contrast results if the minimum shear distance of a half-wavelength is selected.

Fluorescence
The specimen is assumed to be evenly illuminated, and to emit incoherent light with an intensity at a given point proportional to the intensity of the exciting light absorbed. Reabsorption of the emitted light by the specimen is ignored.

Scanning (transmitted or fluorescence)
The numerical aperture, glare, focus and aberration of the lens, and the size of the associated pin-hole are specified for both the objective and condenser. The default value for the pin-hole diameters is half a wavelength of light (diffraction limited). Default values for the condenser are those already chosen for the objective, i.e. confocal scanning is assumed unless other values are chosen. In fluorescence scanning an option is provided to define different wavelengths for the exciting and emitted light.

Microdensitometry
Only a one-dimensional (line) scan is possible. The parameters which must be defined are the left ('FLeft') and right ('FRight') borders of the image zone to the scanned, and the width (SW%) of the measuring spot. The minimum width of the measuring spot is a half-wavelength; the maximum width equals the full width of the field, but only odd numbers

of half-wavelengths are accepted, e.g. 3, 5, 31. Superimposed on the image are the integrated absorbance of the defined area, and the mean absorbance (integrated absorbance divided by the number of positions taken by the measuring spot during the scan). If you wish to redefine any of the three measuring parameters without altering the image, press Y, otherwise touch any key to return to the Main Menu.

Escaping from and re-entering the program
If you leave the Zernike program to run the Kohler, Nicol or Snellius program, and then return to the Zernike program, previously defined values are 'remembered'. If you are running the programs in a Windows 95 or Windows 98 window, you can temporarily leave the program you are in (minimize the window) by pressing ALT together with ESC, and then re-enter the program by clicking on the RUN or ZERNIKE icon at the bottom of the screen.

16.7 USING PARTS OF THE PROGRAM IN AN ARBITRARY ORDER

A feature of interpreted BBC BASIC is its ability to 'call' named procedures and functions in a program directly from the keyboard when the program is loaded but not running. Procedures and functions can be called individually or in a series linked by colons. This is not possible in the compiled IBM version of Zernike, but in the DIY facility (accessed in the program by pressing function key F10) you can specify an action or series of actions which are executed in sequence. These actions include defining the object and altering the instrumental settings (the Main Menu), carrying out a FFT on the data currently in memory, modifying the data either directly or via the coherent imaging mode, and displaying the current data graphically or numerically on the VDU. It is also possible to carry out calculations on the keyboard as if on a calculator, using the current values of the most important program variables held in memory. Full instructions are provided on the screen for all these options, and for returning to the main program.

16.8 DEFINING AN OBJECT WITH FINITE THICKNESS

Although the program inherently assumes an object of negligible thickness in the z axis, when using axial coherent illumination it is possible to simulate an object consisting of a number of discrete and separate layers. First define an object (say a black object of arbitrary width, on a bright background) in the usual way. Carry out the FFT, and using the bright-field imaging option select a numerical aperture of unity, and a defocus of (say) 1 wavelength. Carry out another FFT to compute the Fresnel diffraction pattern at a distance of 1 wavelength from the object. Now manually change values in the arrays (as described above) to define the second layer of the object – say set certain values to zero to simulate another opaque object. You can now study the imaging of this double-layered object by carrying out yet another FFT, modifying the Fraunhofer pattern by using any of the Transmitted-light imaging options, and compute the image by performing a final FFT.

16.9   ACCESSIBLE PROGRAM VARIABLES, ARRAYS AND FUNCTIONS

Table 16-2 shows minimum, maximum and initial default values. The character % in a name indicates an integer (whole number) variable. The computer will accept the name of a variable whenever numerical data are called for – for example, if you want to increase the value of the objective numerical aperture (NA) by the smallest possible amount, type the expression NA + 1/N when defining the value of NA. Names are case-insensitive, so that for example NA and na are equivalent.

Table 16-2   Accessible program variables, arrays and functions

Name of variable	Min. Val.	Max. Val.	Initial Val.	Comment
A(n)	n = 1	n = N%		Array with 'real' components of vectors
Aa	0	9	1	Maximum absorbance of apodization plate
Ap	−9	9	1	Maximum retardation of apodization plate in wavelengths
ApodSym%	0	1	1	Apodization symmetrical about axis (1 = yes)?
axis%	9	129	33	Position in array of optic axis (N/2 + 1).
BA	0	9	9	Maximum absorbance of grating
BOPD	−9	9	0	Maximum retardation of grating
BW%	0	N	N	Width of bar of square grating, or leading edge of sawtooth grating, in half-wavelengths
C%	2	N	N	Grating interval in half-wavelengths
Co	−1	1	0	Obliquity of coherent illumination (sine of angle of light relative to optic axis)
DT	0	1	0	Transmittance of central (axial) objective stop
DW	0	0.75	0	Diameter of central objective stop
FLeft	1	N	1	Left margin of area for microdensitometry
FNA(n)				Returns the retardation, in radians, of the light at image or object point n. Divide by (PI*2) to convert to wavelengths.
FNR(n)				Returns the intensity of the light at image or object point n.
FRight	FLeft	N	N	Right margin of area for microdensitometry
FR	1	2	1	Emitted/exciting wavelengths
FS	0	1	1	Ratio of Field Stop/Total Field Width

Name of variable	Min. Val.	Max. Val.	Initial Val.	Comment
G	0	1	0	Glare due to objective
Gc	0	1	0	Glare due to condenser
gratype%	0	2	0	Grating type (0 = square, 1 = sawtooth, 2 = sinusoidal)
Gt	0	1	0	Transmittance of zero order light in 'Gegenfeld' imaging
HALFN%				Half the current value of N% (see below)
HO	−0.5	0.5	0	Offsetting from axis of 'modulation contrast' modulator
Id	0	1	0.25	Phase difference in wavelengths between object and reference beams in interference contrast
Ir	0	1	0.5	Fraction of total energy in reference beam in interference contrast
Is%	1	n−1	0	Interference shear in half-wavelengths
N%	8	256	64	Number of elements in use in arrays A(n) and P(n), i.e. the width of the field in half-wavelengths
N				Alternative form of N%
n				Alternative form of N%
NA	2/N	1.0	1.0	Numerical aperture of objective
NAc	0	1.0	1.0	Numerical aperture of condenser
Nob%	1	5	1	Number of discrete objects in field
OFA(n)	−N/2	N/2−1	0	Offsetting of object number n (n = 1 to 5) from optic axis in half-wavelengths
OPD(n)	−9	9	0	Retardation of object number n (n = 1 to 5) in wavelengths
OW(n)	1	N	1	Width of object number n (n = 1 to 5) in half-wavelengths
P(n)	n = 1	n = N%		Array with 'imaginary' components of vectors
PR	−1	1	0.75	Phase plate retardation
PT	0	1	1	Phase plate transmittance
Ptype%	0	0	0	Type of phase plate
PW	0	0.75	0	Phase plate diameter
S%	0	3	1	Screen display mode (off, intensity, absolute amplitude or signed amplitude)
Sc	0.5	N/2	0.5	Scanning spot diameter (lamp side) in wavelengths
SHA(n)	0	2	1	Shape of object n (n − 1 to 5). (0 − square, 1 = prism, 2 = sphere)
S1	−999	999	0	Spherical aberration in wavelengths of objective
S1c	−999	999	0	Spherical aberration in wavelengths of condenser
So	0.5	N/2	0.5	Scanning spot diameter (imaging side) in wavelengths
ST%	0	N−1	0	Lateral shift of grating in half-wavelengths
SW%	1	N−1	1	Width of microdensitometric measuring spot in half-wavelengths (odd vals. only)
T(n)	0	1	0	Transmittance of discrete object number n.
Z	−999	999	0	Wavelengths from best focus of objective
Zc	−999	999	0	Wavelengths from best focus of condenser

Chapter 17

Some Suggested Exercises Using the Zernike Program

The following suggested exercises illustrate a few of the important principles in microscopy which can be investigated using the Zernike program. If possible, you should at some convenient time study real objects with a real microscope, and note to what extent the predictions of the program can be confirmed. First carry out each exercise using the suggested values for the object and the imaging system, and try to interpret the results theoretically. Then substitute other object and/or imaging parameters, and see if you can predict the resulting diffraction patterns and images.

The default value of the field diameter (32 wavelengths) is suitable for most exercises using coherent illumination (condenser aperture NAc = 0). Particularly if your computer is old, computing may however be rather slow with incoherent illumination (NAc = 1) and very slow indeed using partially incoherent illumination (0 < NAc < 1); it may then be advisable to select a smaller field width (Main Menu option 4).

One (or two) bright objects on a black background can be obtained by defining two (or three) discrete objects using the 'substitute' option. The first object defined should be the same width as the field of view with zero transmittance; subsequent objects should be narrower, with a finite transmittance.

To define a transparent object specify a transmittance of 1.0 and a finite retardation (not a whole number of wavelengths if the 'square' shape option has been chosen in the Main Menu, since in this case a whole number of wavelengths retardation is equivalent to zero retardation).

17.1 DIFFRACTION BY AN GRATING – EFFECT OF THE GRATING INTERVAL

Define a 'square' grating with absorbing an interval of (say) 32 half-wavelengths, a bar width of 16 half-wavelengths, and the maximum absorbance of 9 for the bars. The retardation of such effectively opaque bars is immaterial. Note the appearance of the corresponding Fraunhofer diffraction pattern. Select bright-field imaging, and observe the effect on the diffraction pattern and the image of reducing the objective NA progressively from 1.0 to 0.75, 0.5, 0.25 and 0.125. How does the fidelity with which the grating is imaged depend on the number of diffracted orders allowed through by the selected objective aperture? What do you see if only the first-order beams are allowed to pass together with the central (zero-order) beam, and what if only the zero-order light is allowed through?

Repeat the exercise, this time reducing or increasing the grating interval but keeping other variables constant. What is the relationship between the grating interval and the number of orders of diffracted light in the Fraunhofer diffraction pattern? How is the diffraction pattern related to the objective NA necessary to 'resolve' the grating?

17.2 DIFFRACTION BY A TRANSPARENT GRATING

Define a square or sinusoidal grating of interval eight half-wavelengths (bar width four half-wavelengths), zero absorbance, and a finite retardation (say 0.1 wavelengths). With a perfectly transparent object, would you expect to see a Fraunhofer diffraction pattern at the back of the objective? Note carefully the predictions of the program for the amplitude and phase relationships of the light across the diffraction pattern, and study their numerical values (press F2). Can you see any regularity in the phase relationships? How does the pattern differ from that seen with an absorbing object of similar dimensions?

17.3 THE AIRY DISC

An Airy disc is the diffraction-limited image of a point (very small) object, produced by a theoretically perfect lens of finite aperture. Define a bright object of minimum width on a dark background. Using bright-field coherent imaging, observe the effect on the Fraunhofer diffraction pattern and the corresponding image of progressively reducing the objective NA. When the NA is about 0.25, the image is a classical Airy disc.

It is instructive to observe such a disc using successively the intensity, absolute amplitude and signed amplitude screen mode options (function key F7). Where the signed amplitude passes from positive to zero, the intensity must be zero – this may not be evident in an intensity or absolute amplitude plot, since none of the discrete points plotted may happen to fall at the minimum.

17.4   RESOLUTION – EFFECT OF OBJECTIVE NA

Define two bright minimum-width objects on a dark background, separated by a wavelength (e.g. with one object at -1 half-wavelength from the optic axis and the other at $+1$). Study the effect on the diffraction pattern and image of progressively decreasing the objective NA from 1.0 to 0.1. What NA is required to *resolve the specimen*, i.e. recognize from the image that two distinct objects are present in the field?

Re-define the objects so that the distance between them is four wavelengths. What is the effect of this on the Fraunhofer diffraction pattern? What objective NA is now required to resolve the specimen?

How do the Rayleigh and Sparrow criteria for resolution differ?

17.5   RESOLUTION – EFFECT OF INCOHERENT ILLUMINATION

Define two adjacent bright point objects on a black background, as before. Are the objects resolved with coherent illumination and an objective NA of 0.5? Change to incoherent illumination (specify a condenser aperture of 1.0 in the Main Menu), recall the same specimens, and use the same objective NA as before. How does this image differ from the previous one? Which type of illumination gives the best resolution? Can you explain this result theoretically? If you have time, study also the effect of partially coherent illumination (e.g. condenser apertures of 0.25, 0.5 and 0.75).

17.6   RESOLUTION – EFFECT OF OBLIQUE ILLUMINATION

Oblique, coherent illumination can be specified from the main Menu (option 6). What effect does an obliquity of (say) 0.5 have on the diffraction pattern of two points, one wavelength apart? Using a small objective aperture (say NA = 0.25), observe the image produced. What is the smallest objective aperture which gives resolution under these conditions?

17.7 RESOLUTION – EFFECT OF FLUORESCENCE OR CONFOCAL SCANNING

What objective aperture is necessary just to resolve two adjacent absorbing objects using fluorescence, transmitted confocal scanning and fluorescence confocal scanning? Choose initially the default (minimum) values for the objective and condenser 'spot widths' in the confocal scanning. Which method gives the best resolving power for a given objective aperture?

In confocal scanning mode, try varying the widths of the scanning spots. What effect does this have on resolution? Is it essential to have the minimum scanning spot size to achieve the benefits of confocal scanning?

17.8 IMAGE OF AN EXTENDED ABSORBING OBJECT – EFFECT OF COHERENT ILLUMINATION

Define an absorbing object on a bright background, the object having a width of (say) 15 half-wavelengths and zero transmittance. Using a finite objective NA (say 0.75), compare the images obtained using first coherent and then incoherent illumination. Note particularly the presence or absence of diffraction fringes near the edges of the specimen, and the minimum apparent transmittance near the centre of the specimen. Which type of illumination gives the most faithful representation of the edges of the object? Which type gives the most accurate reproduction of the central transmittance?

17.9 APODIZATION – OBJECTS OF VARYING SIZE OR TRANSMITTANCE

Define two absorbing objects with the same transmittance (say 0.5) but differing in width; with a field 32 wavelengths wide the objects might be respectively 4 and 8 half-wavelengths wide, situated some distance apart on opposite sides of the optic axis. Study systematically the effect on the images of using symmetrical apodizing stops of varying properties (width, maximum absorbance, maximum retardation, apodization function). Can you find a set of apodization properties which causes the image of the narrower object to appear darker than that of the wider one?

Now re-define the two objects so that they are the same width (say 2 wavelengths), but differ in transmittance (say 0.5 and 0.1 transmittance respectively). Try to use apodization to make the image of the paler object appear darker than that of the more absorbing specimen.

17.10   TRANSPARENT OBJECT – BRIGHT-FIELD IMAGING

Define an object which half fills the field (e.g. 32 half-wavelengths), with a transmittance 1.0 and retardation 0.1. Specify bright-field, in-focus imaging with an objective NA of 1.0. Is a Fraunhofer diffraction pattern visible at the back of the objective? Remembering that only amplitude and not phase properties of an image are visible to the human eye, what image would be seen under these conditions? Reduce the objective NA to 0.5, without changing anything else. Remembering that about 5% intensity modulation is probably necessary for visibility by the human eye, would this image be visible? Now try the effect of throwing the object slightly out of focus (say by 1 wavelength), with an objective NA of first 1.0 and then 0.5. Would these images be visible? Finally, try using incoherent illumination with any combination of objective NA and focus you like. If you have time, study also the effect of partially coherent illumination. What lessons does all this teach us regarding the imaging of a perfectly transparent object with a bright-field microscope? Are analogous effects seen with the transmission electron microscope?

17.11   TRANSPARENT OBJECT – PHASE CONTRAST IMAGING

(a) Specify the same transparent specimen as in the previous exercise, and select phase contrast (option 2 in the coherent imaging menu). Use the default A- type of phase plate with the default transmittance of 1.0, and select a finite objective NA (say 0.75). Observe the effect on the image of gradually increasing the width of the phase plate from the 'ideal' value of zero up to (say) a diameter of 0.25. What is meant by the term 'phase halo'? How is the width of the halo related to the width of the phase plate?

(b) Using a phase plate of zero width and an objective of NA 1.0, note the effect on the image of reducing the transmittance of the phase plate from 1.0 to 0.5, 0.25, 0.125, 0.1 and finally 0.05. Which phase plate gives the darkest image? What is meant by 'reversal of contrast', and under what conditions is this phenomenon observed? Now try a phase plate of transmittance TAN((PI/2)^2), where 0.1 is the defined retardation of the object in wavelengths. What is the intensity of the image now? Try the exercise again, but with an object only (say) one quarter as wide as the whole field. Is the intensity of the image the same as before?

(c) Repeat the exercise using a phase plate of finite radius 1/8, and an objective of NA 0.75. Is the image contrast as good as with ideal phase-contrast? Establish by systematic experiment which features of the image are due to the reduced NA and which to the finite phase plate diameter.

(d) What is the effect of using an A + phase plate, with the same width and transmittance as you found gave good contrast with an A − phase plate? How does the darkness of the centre of the image compare with that of the background? What sort of halo is present?

(e) Absorbing object − effect of phase contrast

(f) With an object 16 wavelengths wide, transmittance (say) 0.5 and retardation zero, study the effect of phase-contrast imaging with a zero-width phase plate of unity transmittance. How much contrast is visible in the image? With what sort of absorbing object might phase-contrast optics be suitable?

17.12 TRANSPARENT OBJECT − OTHER METHODS OF IMAGING

Define a retarding but non-absorbing object of width 16 wavelengths, transmittance 1.0 and retardation 0.1. Try in succession the Schlieren, central dark-field, peripheral dark-field, 'amplitude contrast' and DIC methods of imaging. Draw up a table summarizing the differences and similarities of the various images obtained.

17.13 FRESNEL DIFFRACTION

The Fraunhofer diffraction pattern visible in the back focal plane of the objective is a special case of diffraction, in which the pattern is optically at infinity relative to the diffracting object. Fresnel diffraction refers to the diffraction patterns formed at other distances, for example quite near the object. For the best results in studying Fresnel diffraction first set the field width to the maximum (256 half-wavelengths), then set the illumination to 'Transmitted' (option 3 in the Main Menu) and the illumination to 'Coherent axial' (option 6). Define an object in the usual way (see below for suggestions), and after the computer has automatically carried out a Fourier transform and displayed the corresponding Fraunhofer diffraction pattern, in the Imaging Menu select Bright-field (option 1). Set the objective NA to 1.0 and the focus error to a finite value, say 1 wavelength. The smaller the focus error, the more closely does the Fresnel diffraction pattern resemble the original object. The 'perfect' image now computed is of a plane 1 wavelength away from the object, i.e. the image represents a Fresnel diffraction pattern.

Interesting objects to study include:

(a) a straight edge, e.g. an opaque object half filling the field. If the field width is 256 units, make the opaque object 128 units wide with an offset of 64 units. Note that fringes are present in the bright part of the Fresnel diffraction pattern but not in the dark part – this is quite different from the result obtained using a relatively small objective aperture with the object in focus, when fringes are also present in the darker part of the field. With the original intensity of the illumination at unity, the geometrical edge corresponds to an amplitude of 0.5 in the diffraction pattern, i.e. and intensity of 0.25. See e.g. Jenkins and White (1957, p. 370–371) for calculated and experimental results of a similar object.

(b) Bright slits of varying width. Define two objects, the first opaque and filling the field, and the second with a transmittance of unity and a width between 1 and 35 half-wavelength units. Compare your results with Jenkins and White (1957, p. 374).

(c) An opaque strip, say 5 half-wavelengths wide. Defocussing by (say) 5 wavelengths reveals a Fresnel diffraction pattern with a bright line (seen as a spot in cross-section) in the middle of a dark shadow. Augustin Fresnel presented his theory of diffraction as an entry to a competition organized by the French Academy of Science in 1817. Siméon Poisson deduced the presence of the bright spot in 1818 as a logical consequence and hence a disproof of Fresnel's theory, but the experimental existence of the bright line was shortly afterwards demonstrated by François Arago, the chairman of the prize committee (for fuller accounts of this episode see e.g. Born and Wolf, 1975, p. 375, or Gribbin, 1995). Note that with a small circular aperture the intensity of the central bright spot in the shadow approaches 80% of the original, unobstructed intensity, but the computer program implicitly implies a strip-like obstruction which gives a significantly lower maximum intensity in the centre of the shadow.

(d) An object of minimum width (one half-wavelength). The Fresnel diffraction patterns of this demonstrate Huygen's principle, i.e. the spreading out of a spherical wave from a point (or a cylindrical wave from the line) object. No central dark patch is seen in the Fresnel diffraction pattern, which increases in width as the distance from the object increases. The fact that the outline (envelope) of the diffraction pattern is not always completely smooth is due to the limitations of a discrete Fourier transform with a limited width.

17.14 FRAUNHOFER DIFFRACTION BY A DOUBLE SLIT

This phenomenon, first demonstrated by Young at the beginning of the 19th century, is of great historical importance in optics. Try first two bright point objects (minimum width) on a dark background, separated by (say) 5 half-wavelengths. Examine the corresponding Fraunhofer diffraction pattern. Then try two objects each (say) 5 half-wavelengths wide, separated by a distance between their centres of 15, 20, 25 and finally 30 half-wavelengths. The corresponding Fraunhofer diffraction patterns should closely resemble those calculated and experimentally shown in Jenkins and White (1957, p. 318).

17.15   THE BECKE LINE

The following experiments are best done with a wide field (256 wavelengths). The Becke line phenomenon (Section 3.12.2) is the asymmetrical movement of the diffraction fringes at the edges of transparent objects, illuminated with almost coherent light, as the plane of focus is changed. Define four transparent objects, each (say) 11 half-wavelengths wide. Two of the objects should have a square profile, and two be spherical, with retardations of 0.1 and -0.1 wavelength. Using coherent illumination, bright-field imaging an objective NA of (say) 0.75, note the effect on the images of changing the focus-deviation from zero to 1 or -1 wavelengths. Which sort of object exhibits the most marked diffraction fringes at best focus? Which shows the best Becke line effect on changing focus (i.e. with which sort of object is the difference between the above- and below-focus images the greatest)? Now try the effect of having the retardation of the objects precisely 0.5 or -0.5 wavelengths. Is the Becke line effect still seen with both square and round objects? Is the Becke line effect seen with opaque objects (transmittance zero)? Try repeating the experiments first with partially coherent illumination (condenser NA of about 0.25) and then with incoherent illumination (condenser NA of unity). Is the Becke line effect still seen?

Bibliography

Abbe, E. (1873). Beiträge zur Theorie des Mikroskops und der mikroskopischen Wahrnehmung. *Arch. mikrosk. Anat.* **9**, 413–468.

Abbe, E. (1904). *Gesammelte Abhandlungen,* 2 vols., ed. S.Czapski, Gustav Fischer Verlag, Jena.

Abramowitz, M. (1983). Rheinberg illumination. *International Laboratory,* October 1983, 76–80.

Airy, G.B. (1838). *Camb. Phil. Trans.* **6**, 379. Quoted by Martin (1966).

Baker, H. (1785). *Of Microscopes, and the Discoveries made Thereby,* 2 vols. J. Dodsley, London.

Barer, R. (1952). A vector theory of phase contrast and interference contrast.
    I.   Positive phase contrast. *Journal of the Royal Microscopical Society,* **72,** 10–30;
    II. Positive phase contrast (continued), *Ibid.,* **72,** 81–98;
    III. Negative phase contrast, *Ibid.,* **73,** 30–39;
    IV. B-type phase contrast, *Ibid.,* **73,** 206–215.

Becke, F. (1893). Ueber die Bestimmbarkeit der Gesteinsgemengtheile, besonders der Plagioklase auf Grund ihres Lichtbrechungsvermoegens. *Sitzungsberichte der Akademie der Wissenschaften in Wien,* **102,** part 1, 358–376.

Bedi, C.S. and Goldstein, D.J. (1974). Cytophotometric factors causing apparent differences between Feulgen-DNA contents of different leukocyte types. *Nature* (London), **252,** 439–440.

Bedi, C.S. and Goldstein, D.J. (1976). Apparent anomalies in nuclear Feulgen–DNA contents. Role of systematic microdensitometric errors. *Journal of Cell Biology,* **71,** 68–88.

Benedek, G. (1965). Microscopic study of blood cells from poultry and rabbit using 3D condenser. *Acta Veterinaria,* **15,** 395–404.

Bennett, A.H., Jupnik, H., Osterberg, H. and Richards, O.W. (1946). Phase microscopy. *Transactions of American Microscopical Society,* **65,** 99–131.

Bennett, H.S. (1950): The microscopical investigation of biological materials with polarized light. In: *McClung's Handbook of Microscopical Technique,* Ed. R.McClung Jones, 3rd Edn. Cassell, London)

Beyer, H. (1953). Untersuchungen über den Einfluss der Gestalt der Aperturblende auf die mikroskopische Abbildung beim Phasenkontrastverfahren. *Jenaer Jahrbuch*, 162–209.

Birch, K.G. (1968). A spatial frequency filter to remove zero frequency. *Optica Acta*, **15**, 112–127.

Blout, E.R., Bird, G.R. and Grey, D.S. (1950). Intra-red microspectroscopy. *Journal of the Optical Society of America*, **40**, 304–313.

Born, M. and Wolf, E. (1975): *Principles of Optics*. 5th Ed., Pergamon Press, Oxford.

Bradbury, S. (1989). *Introduction to the Optical Microscope*. Bios Scientific Publishers, Oxford, and Royal Microscopical Society, Microscopy Handbooks No. 01.

Bradbury, S. and Evennett, P.J. (1996). *Contrast Techniques in Light Microscopy*. Bios Scientific Publishers, Oxford, and Royal Microscopical Society, Microscopy Handbooks. No. 34.

Bradbury, S., Evennett, P.J., Haselmann, H. and Piller, H. (1989). *Dictionary of light microscopy*. Oxford University Press and Royal Microscopical Society, Microscopy Handbooks No. 15.

Bratuscheck, K. (1892). Die Lichtstärke-Änderungen nach verschiedenen Schwingungsrichtungen in Linsensystemen von grossem Öfnungswinkel mit Beziehung zur mikroskopischen Abbildung. *Ztschr. f. wiss. Mikr.*, **9**, 145–160.

Bretschneider, F. and Teunis, P.F.M. (1994). Reduced-carrier single-sideband microscopy: a powerful method for the observation of transparent microscopical objects. *Journal of Microscopy*, **175**, 121–134.

Brown, K.M. and McCrone, W.C. (1963). Dispersion staining. Part I – Theory, method and apparatus. *The Microscope and Crystal Front*, **13**, 311–322.

Cannell, M.B. and Soeller, C. (1997). High resolution imaging using confocal and two-photon molecular excitation microscopy. *Proceedings of the Royal Microscopical Society*, **32**, 3–8.

Carpenter, W.B. and Dallinger, W.H. (1891). *The Microscope and its Revelations*, 7th edn. J. and A. Churchill, London , p. 1099.

Caspersson, T. (1936). Über den chemischen Aufbau der Strukturen des Zellkernes. *Skand. Arch. Physiol.*, **73** (Suppl. 8), 1–151.

Chamot, E.M. and Mason, C.W. (1958). *Handbook of chemical microscopy*, vol. **1**, 3rd edn. Chapman and Hall, London.

Conrady, A.E. (1905). An experimental proof of phase-reversal in diffraction-spectra. *Journal of the Royal Microscopical Society*, 150–152.

Cox, I.J., Sheppard, C.J.R. and Wilson, T. (1982). Super-resolution by confocal fluorescent microscopy. *Optik*, **60**, 391–396.

Cox, I.J. and Sheppard, C.J.R. (1986). Information capacity and resolution in an optical system. *Journal of the Optical Society of America* A, **3**, 1152–1158.

Crossmon, G.C. (1948). 'Optical staining' of tissue. *Journal of the Optical Society of America*, **38**, 417.

Crossmon, G.C. (1949). The 'dispersion staining' method for the selective coloration of tissue. *Stain Technology*, **24**, 61–65.

Davidson, J.A. and Butler, R.S. (1992). Sizing particles by optical microscopy through criteria developed by the analysis of image profiles. *Part. Part. Syst. Charact.*, **9**, 213–222.

Dekkers, N.H. and de Lang, H. (1974). Differential phase contrast in a STEM. *Optik*, **41**, 452–456.

Denk, W., Strickler, J.H. and Webb, W.W. (1990). Two-photon laser scanning fluorescence microscopy. *Science*, **248**, 73–76.

Dippel, L. (1872). *Das Mikroskop und seine Anwendung*. Friedrich Vieweg und Sohn, Braunschweig. p. 490.

Dixon, A. (1997). Two-photon fluorescence: a new dimension for microscopy. *International Laboratory*, **27**, 13–16.

Exner, S. (1885). Ein Mikro-Refractometer. *Archiv für mikroskopische Anatomie*, **25**, 97–112.

Exner, S. (1887). Remarks on our knowledge of the structure of the transversally striated muscle–fibres. *Arch. f. d. gesammt. Physiol. (Pflüger)*, **40**, 360–393. Summarized and translated in *Journal of the Royal Microscopical Society*, 1888, pt. 1, 119–135.

Faust, R.C. (1951). Fresnel diffraction at a transparent lamina. *Proc. Phys. Soc. B.*, **64, **105–113.

Faust, R.C. (1955). Refractive index determinations by the central illumination (Becke line) method. *Proc. Phys. Soc.*, **68**, 1081–1094.

Fay, F.S., Carrington, W. and Fogarty, K.E. (1989). 3-D molecular distribution – digitial imaging. *Journal of Microscopy*, **153**, 133–149.

Foucault (1858). Mémoire sur la construction des télescopes en verre argenté. *Annales de l'observatoire imp. de Paris*, Tome V. Cited by Witting (1906).

Frey, H. (1872). *The Microscope and Microscopical Technology*. Translated from the 4th and last German edition by G.R. Cutter. William Wood and Co., New York.

Gage, S.H. (1920). Modern dark-field microscopy and the history of its development. *Transactions of the American Microscopical Society*, **39**, 95–141.

Gage, S.H. (1925). *The microscope. Dark-field (14th) Edition*. Comstock Publishing Co., N.Y.

Gahm, J. (1964). Quantitative polarisationsoptische Messungen mit Kompensatoren. *Zeiss-Mitteilungen* **3**, 153–192.

Galbraith, W. (1982). The image of a point of light in differential interference microscopy: computer simulation. *Microscopica Acta*, **85**, 233–254.

Galbraith, W. and Sanderson, R.J. (1980). The energy distribution about the image of a point. *Microscopica Acta,* **83**, 395–402.

Garcia, A.M. (1965). A one-wavelength, two-area method in microspectrophotometry for pure amplitude objects. *Journal of Histochemistry and Cytochemistry*, **13**, 161–167.

Garcia, A.M. and Iorio, R. (1966). Potential sources of error in two-wavelength cytophotometry. In: *Introduction to Quantitative Cytochemistry*, ed. G.L. Wied. Academic Press, New York, pp. 215–237.

Goldstein, D.J. (1968). New method of thickness measurement with the interference microscope. *Nature*, **213**, 386–387.

Goldstein, D.J. (1969a). The fluorescence of elastic fibres stained with Eosin and excited by visible light. *Histochemical Journal,* **1**, 187–198.

Goldstein, D.J. (1969b). Detection of dichroism with the microscope. *Journal of Microscopy* **89**, 19–36.

Goldstein, D.J. (1970a). Analysis of polarized light with two quarter-wave plates. *Journal of Microscopy*, **91**, 19–30.

Goldstein, D.J. (1970b). Aspects of scanning microdensitometry. I. Stray light (glare). *Journal of Microscopy*, **92**, 1–16.

Goldstein, D.J. (1971). Aspects of scanning microdensitometry. II. Spot size, focus and resolution. *Journal of Microscopy*, **93**, 15–42.

Goldstein, D.J. (1975). Aspects of scanning microdensitometry. III. The monochromator system. *Journal of Microscopy*, **105**, 33–56.

Goldstein, D.J. (1981). Errors in microdensitometry. *Histochemical Journal*, **13**, 251–267.

Goldstein, D.J. (1982a). A simple quantitative analysis of phase contrast microscopy, not restricted to objects of very low retardation. *Journal of Microscopy*, **128**, 33–47.

Goldstein, D.J. (1982b) Scanning microdensitometry of objects small relative to the wavelength of light. *Journal of Histochemistry and Cytochemistry*, **30**, 1040–1050.

Goldstein, D.J. (1986). Random signal variation as a cause of systematic microdensitometric error. *Cytometry*, **7**, 532–535.

Goldstein, D.J. (1991a). A quantitative computer simulation of microscopic imaging. *J. Microscopy*, **162**, 241–253.

Goldstein, D.J. (1991b). Quantitative theory of ideal phase-contrast microscopy, taking object width into account. *Journal of Microscopy*, **164**, 127–142.

Goldstein, D.J. (1992). Resolution in light microscopy studied by computer simulation. *Journal of Microscopy*, **166**, 185–197.

Goldstein, D.J. (1994). Problems and precautions in high-precision two-beam microinterferometry. *Proceedings SPIE (International Society for Optical Engineering)*, **1846**, 244–251.

Goldstein, D.J. and Williams, M.A. (1974). Quantitative assessment of autoradiographs by photometric reflectance microscopy. An improved method using polarized light. *Histochemical Journal*, **6**, 223–230.

Goranson, R.W. and Adams, L.H. (1933): A method for the precise measurement of optical path-difference, especially in stressed glass. *Journal of the Franklin Institute* **216**, 475–504.

Gordon, J.W. (1906). Dark field illumination. *Journal of the Royal Microscopical Society*, 157–160.

Gordon, J.W. (1907). The use of a top stop for developing latent powers of the microscope. *Journal of the Royal Microscopical Society*, 1–13.

Grabham, G.W. (1910). An improved form of petrological microscope. *Min. Mag.*, **15**, 335–349.

Gribbin, J. (1995). *Schrödinger's Kittens and the Search for Reality*. Weidenfeld and Nicholson, London.

Hallimond, A.F. (1947). Production of contrast in the microscopic image by means of opaque diaphragms. *Nature*, **159**, 851–852.

Hallimond, A.F. (1953). *Manual of the Polarizing Microscope*, 2nd edn. Cooke, Troughton and Simms, York.

Hallimond, A.F. (1970). *The Polarizing Microscope*, 3rd edn. Vickers Instruments, York.

Hamilton, D.K. and Wilson, T. (1984). Edge enhancement in scanning optical microscopy by differential detection. *J. Opt. Soc. Amer. Ser. A*, **1**, 322–323.

Hartshorne, N.H. and Stuart, A. (1970). *Crystals and the polarising microscope*, 4th edn. Edward Arnold, London.

Heavens, O.S. (1955). *Optical Properties of Thin Solid Films*. Butterworths, London.

Hegedus, Z.S. (1990). Pupil filters in confocal imaging. Chap. 5 in: *Confocal Microscopy*, ed. T. Wilson. Academic Press, London.

Hiskey, C.F. (1955). Absorption spectroscopy. In *Physical Techniques in Biological Research*, ed. G. Oster and A.W. Pollister. Academic Press, New York, vol. **1**, pp. 215–239.

Hoffman, R. (1977). The modulation contrast microscope: principles and performance. Pt. 3. *Journal of Microscopy*, **110**, 205–222.

Hoffman, R. (1988). Applications of the modulation contrast microscope. *International Laboratory*, **18**, 32–39.

Hoffman, R. and Gross, L. (1975a). The modulation contrast microscope. *Nature*, **254**, 586–588.

Hoffman, R. and Gross, L. (1975b). Modulation contrast microscope. *Applied Optics*, **14**, 1169–1176.

Hogg, J. (1883). *The Microscope: Its History, Construction, and Application*. 10th ed., George Routledge and Sons, London.

Hooke, R. (1665). *Micrographia*. Royal Society, London. Reprinted by Dover Publications, N.Y., 1961.

Hopkins, H.H. and Barham, P.M. (1950). The influence of the condenser on microscopic resolution. *Proceedings of the Physical Society*, **63**, 737–744.

Horikawa, Y., Yamamoto, M. and Shinichi, D. (1987). Laser scanning microscope: differential phase images. *Journal of Microscopy*, **148**, 1–10.

Hotchkiss (1910). Quoted by Johannsen (1914).

Inoué, S. (1986). An introduction to biological polarization microscopy. In: *Video Microscopy*, Plenum Press, New York.

Inoué, S. and Hyde, W.L. (1957). Studies on depolarization of light at microscope lens surfaces. II. The simultaneous realization of high resolution and high sensitivity with the polarizing microscope. *Journal of Biophysical and Biochemical Cytology* **3**, 831–837.

Jacquinot (1950) Cited by Jacquinot and Roizen-Dossier (1964).

Jacquinot, P. and Roizen-Dossier, B. (1964). Apodization. In: *Progress in Optics*, vol. **3**, ed. E. Wolf, North Holland Publishing Co., Amsterdam, pp 29–186.

James, V. and Goldstein, D.J. (1974). Haemoglobin content of individual erythrocytes in normal and abnormal blood. *British Journal of Haematology*, **28**, 89–101.

Jenkins and White (1957). *Fundamentals of Optics*, 3rd edn. McGraw–Hill Book Co. Inc., New York.

Jerrard, H.G. (1948). Optical compensators for measurement of elliptical polarization. *Journal of the Optical Society of America*, **38**, 35–59.

Johannsen, A. (1914). *Manual of Petrographic Methods*. McGraw–Hill, New York. p. 649.

Kachar, B. (1985). Asymmetric illumination contrast: a method of image formation for video light microscopy. *Science*, **227**, 766–768.

Keck, P.H. and Brice, A.T. (1949). Image contrast in phase-contrast microscopy. *Journal of the Optical Society of America*, **39**, 507–514.

Kimura, S. and Munakata, C. (1990). Depth resolution of the fluorescent confocal scanning optical microscope. *Applied Optics*, **29**, 489–494.

Kinder, E. and Recknagel, A. (1947). Ueber Fresnelsche Beugung beim Licht- und Elektronenmikroskop. *Optik, Berl.*, **2**, 346–363.

Köhler, A. (1894). Ein neues Beleuchtungsverfahren für mikrophotographische Zwecke. *Zeitsch. f. wiss. Mikroskopie, 10,* 433–440.

Köhler, A. and Loos, W. (1941). Das Phasenkontrastverfahren und seine Anwendung in der Mikroskopie. *Naturwissenschaften*, **29**, 49–61.

Krakow, W. (1984). Computer simulation and analysis of high-resolution electron microscope images and diffraction patterns with partial coherence, hollow cone illumination, and virtual apertures. *Journal of Electron Microscopy Technique*, **1**, 107–130.

Krakow, W. (1991). Real time computer simulation of transmission electron microscope images with tilted illumination: grain boundary applications. *Journal of Electron Microscopy Technique*, **19**, 366–378.

Lacey, A.J. (1968). Some illustrative material for students of microscopy. *Proceedings of the Royal Microscopical Society*, **3**, 175–185.

Lighton, W. (1878). A new device for dark-field illumination. *American Quarterly Microscopical Journal*, **1**, 42–43.

Lister, J.J. (1830). On some properties in achromatic object glasses applicable to the improvement of the microscope. *Philosophical Transactions of the Royal Society, 120*, 187–200.

Longhurst, R.S. (1973). *Geometrical and Physical Optics*. Longman, London.

Lummer, O. and Reiche, F. (1910). *Die Lehre von der Bildentstehung im Mikroskop von Ernst Abbe*. Friedrich Vieweg und Sohn, Braunshweig.

Macias Garza, F., Diller, K.R., Bovik, A.C., Aggarwal, S.J. and Aggarwal, J.K. (1989). Improvement in the resolution of three-dimensional data sets collected using optical serial sectioning. *Journal of Microscopy*, **153**, 205–221.

Majlof, L. and Forsgren, P. (1993). Confocal microscopy: important considerations for accurate imaging. In: *Cell Biological Applications of Confocal Microscopy*, ed. B. Matsumoto, Vol. 38 of *Methods in Cell Biology*, Academic Press, San Diego, pp. 79–95.

Martin, L.C. (1966). *The Theory of the Microscope*. Blackie, London.

Matsumoto, B. (1993) (Ed). *Cell Biological Applications of Confocal Microscopy*. Vol. 38 of *Methods in Cell Biology*, Academic Press, San Diego, p. 380.

Maschke, O. (1872). Ueber Abscheidung krystallisirter Kieselsäure aus wässerige Lösungen. *Poggendorffs Annalen der Physik*, **144**, 549–578.

McCutchen, C.W. (1967). Superresolution in microscopy and the Abbe resolution limit. *Journal of the Optical Society of America*, **57**, 1190–1192.

McMullan, D. (1990). The prehistory of scanned image microscopy. Part I: Scanned optical microscopes. *Proceedings of the Royal Microscopical Society*, **25**, 127–131.

Mendelsohn, M.L. (1966). Absorption cytophotometry: comparative methodology for heterogenous objects, and the two-wavelength method. In: *Introduction to Quantitative Cytochemistry* ed. G.L. Wied. Academic Press, New York, pp. 202–214.

Menzel, E. (1949). Erhöhter Bildkontrast bei ausgedehnten Objekten. *Optik*, **5**, 385–394.

Menzel, E. (1957). Phasenkontrast-Verfahren. In: *Handbuch der Mikroskopie in der Technik*, ed. H. Freund. Umschau Verlag, Frankfurt a. M., pp. 319–356.

Merkel, F. (1873). Der quergestreifte Muskel II. Der Contractionsvorgang im polarisierten Licht. *Archiv für mikroskopische Anatomie*, **9**, 293–307.

Michel, K. (1964). *Die Grundzüge der Theorie des Mikroskops*. 2nd edn. Wissenschaftliche Verlagsgesellschaft, Stuttgart.

Misell, D.L., Burge, R.E. and Greenaway, A.H. (1974). Alternative to holography for determining phase from image intensity measurements in optics. *Nature*, **247**, 401–402.

Naegeli, C. and Schwendener, S. (1887). *The Microscope in Theory and Practice*. Swan Sonnenschein, Lowrey and Co., London. p. 382. English translation by F. Crisp of 1877 Edition of *Das Mikroskop*.

Naora, H. (1951). Microspectrophotometry and cytochemical analysis of nucleic acids. *Science*, **114**, 279–280.

Narath, A. (1948). Zur Erklärung des Phasenkontrastverfahrens. *Optik*, **4**, 9–10.

Nelson (1910). Cited by Volkmann (1928).

Niemi, M. (1958). Cytophotometry by silver analysis of photomicrographs. *Acta Anatomica*, **35**, 1–92.

Oettlé, A.G. (1950). Experiments with a variable amplitude and phase microscope. *Journal of the Royal Microscopical Society*, **70**, 232–254.

Oster, G. (1955): Birefringence and dichroism. In: *Physical Techniques in Biological Research*, vol. 1, ed. G. Oster and A.W. Pollister. Academic Press, New York.

Picht, J. (1936). Bemerkungen über den Phasenunterschied im Bilde der Fraunhoferschen Beugungserscheinungen. II. Zum Phasenkontrastverfahren von Zernike. *Zeitschrift für Instrumentenkunde*, **12**, 481–489.

Pluta, M. (1975). Non-standard methods of phase contrast microscopy. In: *Optical and Electron Microscopy*, ed. R. Barer and V.E. Cosslett, vol. **6**, Academic Press, London, pp. 49–133.

Pluta, M. (1989). Simplified polanret system for microscopy. *Applied Optics*, **28**, 1453–1466.

Pluta, M. (1988–89). *Advanced Light Microscopy*. 3 vols., Elsevier, Amsterdam.

Quekett, J. (1852). *Practical Treatise on the Use of the Microscope*. 2nd edn. H. Bailliere, London.

Raveau (1902). Quoted by Witting (1906).

Rayleigh, Lord (1896). On the theory of optical images, with special reference to the microscope. *Phil. Mag. Series 5*, **42**, 167–195.

Reade, J.B. (1837). A new method of illuminating microscopic objects. In: *'Micrographia'*, C.R. Goring and A. Pritchard. Whitaker and Co., London, pp. 227–231.

Reynolds, G.O., DeVelis, J.B., Parrent, G.B. and Thompson, B.J. (1989). *The New Physical Optics Notebook: Tutorials in Fourier Optics*. SPIE Optical Engineering Press, Bellingham.

Rheinberg, J. (1905). The influence on images of gratings of phase-difference amongst their spectra. *Journal of the Royal Microscopical Society*, 152–155.

Richartz, M. and Hsü, H.Y. (1949). Analysis of elliptical polarization. *Journal of the Optical Society of America* **39**, 136–157.

Richter, R. (1947). Eine einfache Erklärung des Phasenkontrastmikrosops. *Optik*, **2**, 342–345.

Rienitz, J. (1975). Schlieren experiment 300 years ago. *Nature*, **254**, 293–295.

Robinson, P. and Bradbury, S. (1992). *Qualitative polarized light microscopy.* Royal Microscopical Society, Oxford Microscopy Handbooks No. 9.

Ross (1855). Cited by Siedentopf (1907b).

Salomon, W. (1896). Ueber die Berechnung des variablen Werthes der Lichtbrechung in beliebig orientirted Schnitten optisch einaxiger Mineralien von bekannter Licht- und Doppelbrechung. *Zeitschr. f. Kryst.*, **26**, 182.

Sarafis, V. (1990). Biological perspectives of confocal microscopy. Chap. 12 in: *Confocal Microscopy*, ed. T. Wilson. Academic Press, London.

Sawyer, D.W., Sullivan, J.A. and Mandell, G.L. (1985). Intracellular free calcium localization in neutrophils during phagocytosis. *Science*, **230**, 663–666.

Saylor, C.P. (1935). Accuracy of microscopical methods for determining refractive index by immersion. *Bur. Stand. J. Res. Wash.*, **15**, 277–94.

Saylor, C.P. (1966). Accurate microscopical determination of optical properties on one small crystal. In: *Advances in Optical and Electron Microscopy*, ed. R. Barer and V.E. Cosslett. Academic Press, London. vol **1**, pp. 41–76.

Schroeder van der Kolk, J.L.C. (1892). *Zeitschr. wiss. Mikroskopie*, **viii**, 456–458. Cited by Wright (1913).

Schroeder van der Kolk, J.L.C. (1898). *Kurze Anleitung zur mikroskopischen Kristallbestimmung.* Wiesbaden. Cited by Wright (1913).

Seifriz, W. (1931). The Spierer lens and what it reveals in cellulose and protoplasm. *Journal of Physical Chemistry*, **35**, 118–129.

Seifriz, W. (1936). Spierer lens and colloidal structure. *Journal of industrial Engineering Chemistry*, **28**, 136–140.

Shadboldt, G. (1851). *Transactions of Microscopical Society London* III, **132**, 154. Cited by Siedentopf (1907).

Sheppard, C. (1987). Scanning optical microscopy. In: *Advances in Optical and Electron Microscopy*, ed. R. Barer and V.E. Cosslett. Academic Press, London, vol. **10**, 1–98.

Sheppard, C.J.R. and Cogswell, C.J. (1990). Three-dimensional imaging in confocal microscopy. Chap. 5 in: *Confocal Microscopy*, ed. T. Wilson. Academic Press, London.

Shuman, H. (1988). Contrast in confocal scanning microscopy with a finite detector. *Journal of Microscopy*, **149**, 67–71.

Shurcliff, W.A. (1962). *Polarized light.* Oxford University Press.

Siedentopf, H. (1907a). Paraboloid-Kondensor, eine neue Methode für Dunkelfeldbeleuchtung zur Sichtbarmachung und zur Moment-Mikrophotographie lebender Bakterien etc. (insbesonder auch für *Spirochaete pallida*). *Zeitschrift für wissenschaftliche Mikroskopie*, **24**, 104–108.

Siedentopf, H. (1907b). Die Vorgeschichte der Spiegelkondensoren. *Zeitschrift für wissenschaftliche Mikroskopie*, **24**, 382–395.

Siedentopf, H. (1909). Über ultramikroskopische Abbildung. *Zeitschrift für wissenschaftliche Mikroskopie*, **26**, 391–410.

Siedentopf, H. (1915). Über das Auflösungvermögen der Mikroskope bei Hellfeld-

und Dunkelfeldbeleuchtung. *Zeitschrift für wissenschaftliche Mikroskopie*, **32**, 1–42.

Spangenberg, K. (1921). Erscheinungen an der Grenze von duennen Objekten im Mikroskop. *Ztschr. f. wiss. Mikr.*, **38**, 1–28.

Sparrow, C.M. (1916). On spectroscopic resolving power. *Astrophyical Journal*, **44**, 76–86.

Spierer, C. (1926). Un nouvel ultra-microscope à éclairage bilateral. *Arch. Sci. phys. nat.* **8**, 121–131.

Spitta, E.J. (1909). *Microscopy*. John Murray, London.

Spycher, R., Stadelmann, P., Buffat P. and Flueli, M. (1988). High-resolution electron microscopy image simulation of a Cray 1S/2300 computer. *Journal of Electron Microscopy Technique*, **10**, 369–372.

Steel, W.H. and Tchan, Y.T. (1960). On the interpretation of phase-contrast images. *Journal of Microscopical Science*, **101**, 465–473.

Stephenson, J.W. (1878). Note on the effect produced on *P. angulatum* and other test objects of excluding the central dioptric beam of light. *Journal of the Royal Microscopical Society*, **1**, 186–187.

Taylor, R.B. (1984). Rheinberg updated. *Proceedings of the Royal Microscopical Society*, **19**, 253–256.

Tonkelaar, E.M. den, and van Duijn, P. (1964). Photographica colorimetry as a quantitative cytochemical method. I. Principles and practice of the method. *Histochemie*, **4**, 1–9.

Töpler, A. (1864). *Beobachtungen nach einer neuen optischen Methode*. Max Cohen und Sohn, Bonn.

Töpler, A. (1866). Über die Methode der Schlierenbeobachtung als mikroskopisches Hilfsmittel, nebst Bemerkungen zur Theorie der Schiefen Beleuchtung. *Poggendorff's Annalen der Physik*, **127**, 556–579.

Töpler, A. (1882). *Ztschr. f. Instrumentenk.* ii, 92–96. Abstracted in *Journal of the Royal Microscopical Society*, Ser. II, vol. V, part 4, 710–713 (1885).

Töpler, A. (1906). Beobachtungen nach der Schlierenmethode. Reprint edited by A. Witting, Wilhelm Engelmann Verlag, Leipzig.

Toraldo di Francia, G. (1955). Resolving power and information. *Journal of the Optical Society of America*, **45**, 497–501.

Uber, F.M. (1939). Ultra-violet spectrophotometry of *Zea mays* pollen with the quartz microscope. *American Journal of Botany*, **26**, 799–807.

Volkmann, R. v. (1928). Über die Verwendung von Zentralblenden zur Verbesserung des mikroskopischen Bildes. *Zeitschrift für wissenschaftliche Mikroskopie*, **45**, 374–379.

Welford, W.T. (1972). On the relationship between the modes of image formation in scanning microscopy and conventional microscopy. *Journal of Microscopy*, **96**, 105–107.

Wenham, F.H. (1878). On the formation of the paraboloid as an illuminator for the microscope. *American Quarterly Microscopical Journal*, **1**, 186–191.

White, J.G., Amos, W.B. and Fordham, M. (1987). An evaluation of confocal versus conventional imaging of biological structures by fluorescence light microscopy. *Journal of Cell Biology*, **105**, 41–48.

Wilson, T. (1990a). Confocal microscopy. In: *Confocal Microscopy* ed. T. Wilson, Academic Press, London, pp. 1–64.

Wilson, T. (1990b). Optical aspects of confocal microscopy. In: *Confocal Microscopy*, ed. T. Wilson, Academic Press, London, pp. 93–141.

Wilson, T. and Carlini, A.R. (1988). Three-dimensional imaging in confocal imaging systems with finite sized detectors. *Journal of Microscopy*, **149**, 51–66.

Wilson, T. and Sheppard, C.J. R. (1981). The halo effect of image processing by spatial frequency filtering. *Optik*, **9**, 19–23.

Wilson, T. and Sheppard, C. (1984). *Theory and Practice of Scanning Optical Microscopy*. Academic Press, New York.

Witting (1906). Editor of and commentator on reprint of Töpler's work in *Ostwalds Klassiker der exakten Naturwissenschaften* No. 158.

Wolter, H. (1950a). Experimentelle und theoretische Untersuchungen zur Abbildung nichtabsorbierender Objekte. *Annalen der Physik Ser. 6*, **7**, 33–53.

Wolter, H. (1950b). Zur Abbildung zylindrischer Phasenobjekte elliptischen Querschnitts. *Annalen der Physik Ser. 6*, **7**, 147–156.

Wolter, H. (1950c). Verbesserung der abbildenden Schlierenverfahren durch Minimumstrahlkennzeichnung. *Annalen der Physik Ser. 6*, **7**, 182–192.

Wolter, H. (1950d). Schlieren–, Phasenkontrast– und Lichtschnittverfahren. In: *Handbuch der Physik* vol. 24, ed. S. Flügge, pp. 555–645.

Woodward, J.J. (1878). Further remarks on a 'simple device' for the illumination of balsam-mounted objects for examination with immersion objectives whose balsam angle is 90° or upwards. *Journal of the Royal Microscopical Society*, **1**, 246–248.

Wright, F.E. (1913). Oblique illumination in petrographic microscope work. *American Journal of Science* Ser. 4, **35**, 63–82.

Yamamoto, K. and Taira, A. (1983). Some improvements in the phase contrast microscope. *Journal of Microscopy*, **29**, 49–62.

Young, M. (1983). Linewidth measurement by high-pass filtering: a new look. *Applied Optics*, **22**, 2022–2025.

Young, M. (1989). Spatial filtering microscope for linewidth measurements. *Applied Optics*, **28**, 1467–1473.

Zernike, F. (1934). Beugungstheorie des Schneidenverfahrens und seiner verbesserten Form, der Phasenkontrastmethode. *Physica, 's Gravenhage*, **10**, 689–704.

Zernike, F. (1935). Das Phasenkontrastverfahren bei der mikroskopischen Beobachtung. *Physikalishe Zeitschrift*, **36**, 848–851.

Zernike, F. (1942). Phase contrast, a new method for the microscopic observation of transparent objects. *Physica, 's Gravenhage*, **9**, 686–698 and 974–986.

Zernike, F. (1955). How I discovered phase contrast. *Science*, **121**, 345–349.

Zselyonka, L. and Kiss, F. (1961). Der dreidimensional (Super-) Kondensor. *Mikroskopie*, **15**, 3–23.

Index